COMPUTER ASSISTED EXERCISES AND TRAINING

COMPUTER ASSISTED EXERCISES AND TRAINING

A Reference Guide

ERDAL ÇAYIRCI
University of Stavanger

DUSAN MARINCIC
NATO Joint Warfare Center

WILEY

A JOHN WILEY & SONS, INC., PUBLICATION

Disclaimer

Definitions, views, and opinions expressed in this book do not necessarily reflect the official point of view of any national or international organization. The content of this book is either compiled from open and unclassified material available also in the Internet or reflects the personal view and opinion of the authors on generic requirements and architectures not tailored for a national or an international organization.

Published by John Wiley & Sons, Inc., Hoboken, New Jersey
Published simultaneously in Canada

For general information on our other products and services or for technical support, please contact our Customer Care Department within the United States at (800) 762-2974, outside the United States at (317) 572-3993 or fax (317) 572-4002.

Wiley also publishes its books in a variety of electronic formats. Some content that appears in print may not be available in electronic formats. For more information about Wiley products, visit our web site at www.wiley.com.

Library of Congress Cataloging-in-Publication Data:

Çayirci, E. (Erdal)
Computer assisted exercises & training : a reference guide / Erdal Çayirci, Dusan Marincic.
p. cm.
"Published simultaneously in Canada"–T. p. verso.
Includes bibliographical references and index.
ISBN 978-0-470-41229-9 (cloth)
1. Military education. 2. Drill and minor tactics–Computer-assisted instruction. 3. Military maneuvers–Computer-assisted instruction. 4. Drill and minor tactics–Computer simulation. 5. Military maneuvers–Computer simulation. 6. Computer war games. I. Marincic, Dusan. II. Title. III. Title: Computer assisted exercises and training.
U405.C39 2009
355.5078′5–dc22 2009011949

10 9 8 7 6 5 4 3 2 1

To Tülin, Ertuğ, Cemre, Lara, and Tuana
Erdal Çayırcı

To Jelka, Andrea, and Sebastian
Dušan Marinčič

CONTENTS

PREFACE xiii

ABOUT AUTHORS xv

PART I FUNDAMENTALS AND THEORY 1

1 Introduction 3

1.1 Contemporary Security Environment 3

1.2 Exercises 7

1.2.1 Military and Civilian Exercises 8

1.2.2 Live Exercises 9

1.2.3 Command Post Exercises 9

1.2.4 Computer-Assisted Exercises 10

1.3 Military Simulation 12

1.4 Scope of the Book 14

1.5 Structure of the Book 15

1.6 Electronic Resources for the Book 17

1.7 Review Questions 17

2 Conflict and Warfare 19

2.1 Paradigms of War 21

2.2 Evolution of Warfare 23

2.2.1 First, Second and Third Generation of Warfare 24
2.2.2 Fourth and Fifth Generation of Warfare 25

2.3 Operations 27
2.3.1 Conventional Operations 28
2.3.2 Special Operations 31
2.3.3 Crises Response Operations 33
2.3.4 Peace Operations 35
2.3.5 Network-Centric Warfare 38
2.3.6 Logistics 39
2.3.7 Information Operations 40
2.3.8 Psychological Operations 41
2.3.9 Effect-Based Approach to Operations (EBAOs) 43

2.4 Comprehensive Approach to Operations 44
2.4.1 Civil Military Cooperation 47
2.4.2 Economical and Social Aspects 48
2.4.3 Comprehensive Approach and Its Application 51

2.5 Review Questions 55

3 Statistics and Probability **57**

3.1 Descriptive Statistics: Population, Sample, Central Tendency, and Dispersion 58

3.2 Probability 64
3.2.1 Counting Techniques 66
3.2.2 Independence, Multiplication Rule, and Conditional Probability 70
3.2.3 Mutually Exclusive Events and Addition Rule 73
3.2.4 Total Probability and Bayes' Theorem 74

3.3 Random Variable 76
3.3.1 Discrete Distributions 76
3.3.1.1 The Uniform Distribution 79
3.3.1.2 The Binomial Distribution 79
3.3.1.3 The Geometric Distribution 79
3.3.1.4 The Negative Binomial Distribution 80
3.3.1.5 The Hypergeometric Distribution 81
3.3.1.6 The Poisson Distribution 81
3.3.2 Continuous Distributions 82

3.3.2.1 The Continuous Uniform (Rectangular) Distribution 85
3.3.2.2 The Exponential Distribution 85
3.3.2.3 The Normal (Gaussian) Distribution 86

3.4 Inferential Statistics 89
3.4.1 Confidence Interval 89
3.4.2 Hypothesis Test 92
3.4.3 Goodness of Fit 96

3.5 Review Questions 101

4 Simulation **103**

4.1 Pseudorandom Number Generation and Realization of Random Variables 105
4.1.1 Pseudorandom Number Generation 105
4.1.2 Realization of Random Variables for a Simulation 110

4.2 Static Simulation 114

4.3 Dynamic Simulation 115
4.3.1 Discrete Event Simulation 116
4.3.2 Continuous Simulation 116

4.4 Phases in a Simulation 117

4.5 Review Questions 118

5 Distributed Simulation **121**

5.1 Distributed Interactive Simulation 124

5.2 High-Level Architecture 127
5.2.1 HLA Interface Specification 129
5.2.2 Object Model Template (OMT) 132
5.2.3 FEDEP 137
5.2.4 HLA Rules 139

5.3 Base Object Model (BOM) 140

5.4 Review Questions 140

6 Experimentation and Analysis **143**

6.1 Design of Experiment 144

6.2 Execution of Experiments 147

6.3 Data Analysis, Reporting, and Presentation 150

6.4 Review Questions 152

PART II COMBAT MODELING, COMPUTER-ASSISTED EXERCISES, AND PRACTICE **153**

7 Computer-Assisted Exercise (CAX) Architectures **155**

7.1 Distributed Exercises and Distributed Simulation 155

7.2 Multilevel and Multiresolution Exercises 160

7.3 Cross-Level, Joint, and Combined Exercises 163

7.4 Excon Structure 167

7.5 Response Cells 169

7.6 Training Audience 170

7.7 Review Questions 170

8 CAX PROCESS **171**

8.1 Exercise Specification 171

8.2 Planning and Preparation 174

8.2.1 Scenario Development 176

8.2.2 MEL/MIL Development 177

8.2.3 CAX Databases and Database Management Process 178

8.3 Execution 180

8.4 Analysis 183

8.5 Review Questions 188

9 Combat Modeling **189**

9.1 Terrain Modeling 193

9.1.1 Environmental Data Representation 194

9.1.2 Data Coding Standard (DCS) 197

9.2 Attrition and Movement 201

9.2.1 Lanchester Equations 203

9.2.2 Stochastic Processing 206

9.3 Challenges in the Quantification for Nonkinetic Warfare 207

9.4 Automated Forces 214
9.5 Challenges and Approaches in the Implementation 220
9.5.1 Complex Systems and Fuzzy Trees 220
9.5.2 Minimalist Modeling Methodology 223
9.5.2.1 Structural Simplicity 224
9.5.2.2 Behavioral Simplicity 224
9.5.2.3 Mathematical Elegance 225
9.5.2.4 Statistical Analysis 225
9.5.2.5 Organic Units 226
9.6 Combat Model Data 227
9.6.1 Organizational Data 227
9.6.2 Equipment, Weapons, and Ammunition Data 227
9.6.3 Terrain Data 227
9.6.4 Environmental Data 228
9.7 Verification and Validation of Combat Models 228
9.8 Experimentation and Analysis of Operational Plans 230
9.9 Review Questions 231

10 Computer-Assisted Exercise Support Tools **233**

10.1 Military Constructive Simulations and Ancillary Tools 234
10.1.1 High-Resolution Constructive Simulations 235
10.1.2 Highly Aggregated Constructive Simulations 239
10.1.3 Constructive Simulations for Nonkinetic Warfare 242
10.1.4 Federations 243
10.1.5 Ancillary Tools 245
10.2 Planning and Management Tools 246
10.2.1 Exercise Management Tools 247
10.2.2 Scenario Management Tools 249
10.3 Mediation-Ware 252
10.4 Review Questions 253

11 Communications/Information System Issues, Technical Risks, and Risk Mitigation **255**

11.1 Hardware and Software Requirements 255
11.2 Communications and QoS Requirements 256
11.3 Security Issues and Challenges 260

11.4 Game Crashes, Checkpoints, and Crash Recovery 262

11.5 Shadow/Run Ahead Games 263

11.6 Backups and Archives 264

11.7 Networking Service Outages and Other Reasons for Failure 264

11.8 Review Questions 265

12 Exercise Centers and Facilities **267**

12.1 Organization of a Training/Exercise Center 269

12.1.1 Operational Staff in a Training/Exercise Center 269

12.1.2 Technical Staff in a Training/Exercise Center 272

12.1.3 Support Staff in a Training/Exercise Center 274

12.1.4 Number of Teams and Staff in a Training/Exercise Center 274

12.2 Design Principles for Training/Exercise Center Facilities 276

12.3 Review Questions 280

REFERENCES **281**

ACRONYMS **287**

INDEX **293**

PREFACE

Contemporary security threats, warfare paradigms, composition of headquarters, and the complexity of operations introduce new challenges for the decision-making and operational planning processes and operating procedures of headquarters. Operational headquarters are often composite organizations made up of international military staff augmented by governmental and nongovernmental, national or international, organizations. This fact exacerbates new challenges introduced by the new generation of warfare, which makes the training of headquarters more and more complex. Emerging combat modeling and information technologies offer effective approaches that can tackle the complexities of this task. Therefore, computer-assisted exercises (CAX) aim to immerse the training audience in an environment as realistic as possible and to support exercise planning and control personnel in such a way that they can steer the exercise process toward the exercise objectives as effectively as possible. It has become the main tool for the headquarters' training.

The book is designed as a comprehensive teaching material for a course on computer-assisted exercises. Basic prerequisite knowledge on military operations and exercises is not required but can be helpful. The book is self-contained on the fundamental probability theory and statistics-related issues, and it provides advanced information on military simulations and CAX. The readers of this book are either exercise planners or technical support personnel, who study to plan a CAX or perform CAX support, CAX system design, and implementation tasks.

This textbook is organized for 14–18 week (3 hours a week) courses. It is also aimed to be a reference book for practitioners, i.e., CAX planners and engineers in industry or in military organizations. The book has two parts. The first one introduces fundamentals and key issues related to the military simulation. In the second part, combat modeling, military simulation, CAX planning, and execution-related issues and technologies are elaborated.

A contemporary security environment demands interdisciplinary studies and research. Most examples described in the chapters are a result of thorough research and discoveries made by both authors in the last 10 years. The situational complexity in the areas of complex emergency request well-trained, flexible, knowledgable, and sustainable operational elements from the International Community. To achieve that, it is of utmost importance to provide state-of-the-art educational and training methodology with the pooling of findings from computer science, natural science, science of mathematics, social science, and military science. With the permanent monitoring of security developments in the globalized world, it is possible to establish and maintain common security data bases. Accessible and updated information allows realistic preparation, organization, and execution of functional training for the designated forces, which becomes closer and closer to the real-world crisis response operations.

The secondary objective of this book is to create conditions for a chain of events, from formation of exercise centers, to education and training of planners and technical personnel for CAX, followed by proper preparation of operational elements for certain crisis situations and ending with the crisis response operations. A structured and timely feedback from the field can than improve the overall training process in the future. With this chain of desired events, authors would like to contribute to improve the life of the affected population in the areas of complex emergency. The authors amalgamate technical standards with the societal security discoveries to make the training methodologies compatible and applicable in a real-world crisis. With that said, they would like to make their humble contribution to the global security and assist in ushering in a brighter future for humanity.

ABOUT AUTHORS

Erdal Cayirci graduated from the Army Academy in 1986 and from the Royal Military Academy, Sandhurst in 1989. He received his Master of Science degree from Middle East Technical University in 1995 and a PhD degree from Bogazici University in 2000, both in computer engineering. He retired from the Army when he was a colonel in 2005. He was an associate professor at the Istanbul Technical University, Yeditepe University, and Naval Sciences and Engineering Institute between 2001 and 2005. Also in 2001, he was a visiting researcher for the Broadband and Wireless Networking Laboratory and a visiting lecturer at the School of Electrical and Computer Engineering at the Georgia Institute of Technology. He is currently Chief, CAX Support Branch in NATO's Joint Warlare Center in Stavanger, Norway, and he is a professor in the Electrical and Computer Engineering Department at the University of Stavanger. His research interests include sensor networks, mobile communications, tactical communications, and military constructive simulation.

Professor Cayirci has acted as an editor of the journals *IEEE Transactions on Mobile Computing*, *AdHoc Networks* (Elsevier Science), and *ACM/Kluwer Wireless Networks*, and he has guest edited four special issues of *Computer Networks* (Elsevier Science), *AdHoc Networks* (Elsevier Science), and *Kluwer Journal on Special Topics in Mobile Networking and Applications* (MONET).

He received the "2002 IEEE Communications Society Best Tutorial Paper" Award for his paper titled "A Survey on Sensor Networks" published in the *IEEE Communications Magazine* in August 2002, the "Fikri Gayret" Award from the Turkish Chief of General Staff in 2003, the "Innovation of the Year" Award from the Turkish Navy in 2005, and the "Excellence" Award at ITEC 2006.

Dusan Marincic has been studying constructive simulations since 1995. He was the Head of a National Centre for Operational Research, Simulation, and Analysis in Slovenia from 1999 until 2005. He performed his master studies with the Faculty of Social Science in September 2002 and researched the topic "CAX as a method for preparation of Peace forces for Peace Support Operation". He then followed his passion for computer-assisted exercises by enrolling in doctoral studies in same faculty and successfully defended his doctoral dissertation in May 2005 with the topic "Simulation and Analysis of Peace Operation." Marincic has written more than 70 articles on peace operations, crisis response operations, and the qualitative and quantitative analysis of peace forces efficiency in areas of complex emergency. All papers were published in national and international journals. He has been a permanent professor with the Command and Staff College of Slovenia and has taught a methods of military sciences, course which examines modern educational and scientific methods. He was twice the general chair of International Week of Simulations in 2002 and 2005, where all of his methodological scientific results were presented to the international experts in the spirit of exchanging knowledge and experience in the area of the computer-assisted exercises. He currently serves as a subject matter expert at NATO's Joint Warfare Center in Stavanger, Norway, where he has been involved with the training and education of NATO'S joint force commands.

PART I

FUNDAMENTALS AND THEORY

1

INTRODUCTION

1.1 CONTEMPORARY SECURITY ENVIRONMENT

At the beginning of the 21st century, the global International Community (IC) was dominated by the democratization of information, technologies, and finances. The key causes of this phenomenon are undoubtedly the development and accessibility of information and communication technology as well as databases on the World Wide Web. Despite the comprehensive functional capabilities of computer equipment, it has not been used enough as a tool for evaluation of security threats to the fundamental values and norms in the international community. Domestic conflicts represent most contemporary security crises in the world, and they often cause the disintegration of political balance in a certain region. The United Nations (UN) Charter makes the Security Council of the United Nations responsible to react to security deviations in the world.

From a historical perspective, security has been the fundamental value of human interactions, which was institutionalized by the rise of the sovereign state and the systems of states on the global level. Grizold [GRI99] emphasized the fact that security has been related to the individuals, society/state, and the international system. Contemporary security patterns involve dealing with individual security, national security, international security, and global security. Møller [MOL00] was defining security, in an objective sense, as a certain

Computer Assisted Exercises and Training: A Reference Guide, by Erdal Çayirci and Dusan Marincic

degree of absence of threats to the reached values, and in a subjective sense as an absence of apprehension that these values will be endangered.

Edmonds writes that the best security condition is the prediction of the future events to execute appropriate preparations [EDM88]. Security is comprehending planning and decision making about activities, which will ensure, together with the available sources, solutions for expected and unexpected threats. Individuals and societies have been recognized with the firm confidence that they can deal with all challenges and threats in the future, as safe and secure. The security system provides in general to the individuals, groups, and society; the means to gain the mastery over unexpected and protection of internal values are included.

Viotti asserted that security is more than just a military way of thinking [VIO94]. It is a defense against external or internal threats given the socioeconomical welfare of society and individuals. After the end of the Cold War, the political science field holistically understands security as a whole spectrum of political, economical, medical, environmental, criminal, and military information and other threats to the modern society. Prezelj confirmed the previous statement by writing that the worst crises are a result of the extreme intensification of threats in one societal dimension, which contributes to the intensification of security threats on other societal dimensions, as depicted in Figure 1.1 [PRE02].

So a combination of crises exist in many security dimensions, or a combination of threats exist from the many dimensions. Military engagements of high intensity have a direct or indirect impact on the intensification of criminal activities and terrorism, as well as on environmental, economical, medical, political, identity, information and other security threats, and vice versa. A complex security threat to the society is in this case a threat with a complex cause (from many dimensions) and with complex effects (on other dimensions).

Security threats to contemporary society are not only military but also political, economical, social, demographical and environmental, which represent fundamental societal security dimensions. Since the end of the Cold War, global peace has been interrupted by numerous conflicts that require international intervention and the deployment of military, police, and civil organizations in peace operations to almost every region in the world. At the same time, the media and the public are continuously watching military leaders, politicians, and other decision makers in whatever they do. Not a single day passes by in which the media does not cover events in crisis areas, and often the focus is on the IC in its attempts to improve the overall security situation. Wherever complex emergency operations are employed, the IC's efforts are affected by many internal and external influences, which vary from local politics, media, population demands, different international organizations (IOs), and nongovernmental organizations (NGOs). Right or wrong, the IC sometimes receives blame for its failure to create a stable and secure environment. Examples like Kosovo in the Balkans and more recently Afghanistan and Iraq indicate that despite all efforts by the IC, a permanent solution to the conflict situation has

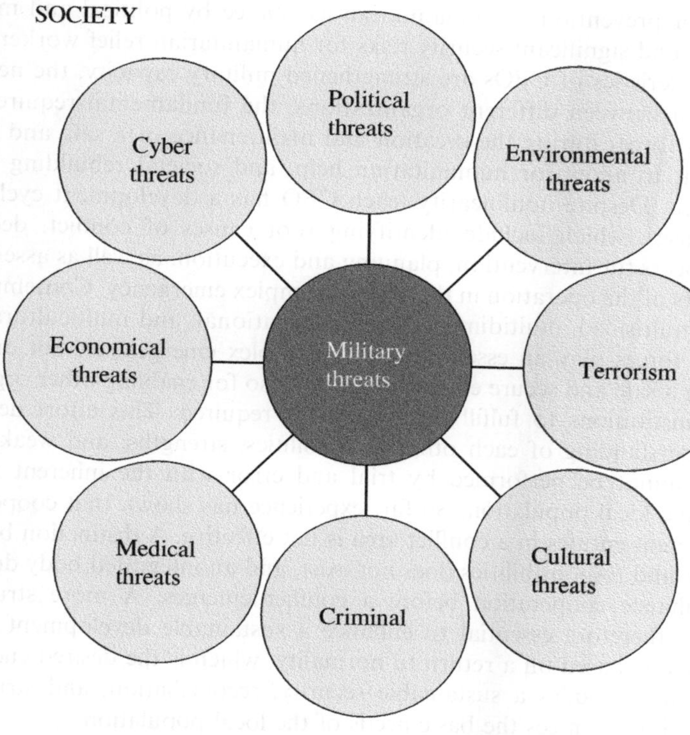

Figure 1.1: Model of the complex security threats to society.

not yet been achieved. Taking into consideration all the different players and the wide variety of agendas, peace and security may never be achieved.

Rebuilding societies is one of the most complex and important challenges the IC faces today. It is absolutely critical to identify areas for improvement to reach the desired end-state of a return to normalcy in a more efficient way and to ensure the future commitment of the IC in assisting with difficult challenges elsewhere. Because of security threats to the main societal security dimensions, complex emergency operations (CEOs) are dynamic nonlinear processes in which the IC, with the help of military forces, rebuilds fundamental societal values and norms.

The official definition of a CEO used in this book is "an operation to address a humanitarian crisis in a country, region or society where there is total or considerable breakdown of authority resulting from internal or external conflict and which requires an international response that goes beyond the mandate or capacity of any single agency and/or the ongoing UN country program." Such complex emergencies are typically characterized by extensive violence and loss of life, massive displacements of people, widespread damage to societies and economies, the need for large-scale humanitarian assistance, the hindrance or prevention of humanitarian assistance by political and military constraints, and significant security risks for humanitarian relief workers. The basic characteristics of CEOs are strengthened military capacity, the need for coordination between different organizations, the fundamental requirements for military forces during the creation and maintenance of a safe and secure environment to allow for humanitarian help, and societal rebuilding of the affected area. Despite nonlinearity, each CEO has a development cycle with specific phases, which include identifying root causes of conflict, decision-making process for intervention, planning and execution, as well as assessment of the effects of the operation in the area of complex emergency. Contemporary CEOs are multisided, multidimensional, multinational, and multicultural.

Military forces play an essential role in complex emergencies, not only for establishing a safe and secure environment but also for enabling other organizations and institutions to fulfill their duties as required. This effort demands mutual understanding of each other's capabilities, strengths, and weaknesses. This task cannot be performed by trial and error with the inherent risk of alienating the local population. So far, experience has shown that cooperation among different entities in a conflict area is not effective. A distinction between roles, tasks, and responsibilities does not exist, and an integrated body does not exist to enhance cooperation before a conflict emerges. A more structured approach is therefore essential to enhance a sustainable development for an affected society, based on a return to normality, which is the desired end state. This normality includes a sustainable security, reconciliation, and structured society, which guarantees the basic needs of the local population.

Military forces are generally essential in the initial stabilization phase, because other organizations might not be ready for deployment or are not functioning effectively yet. A transition to civil authorities, whether international or national, requires a clear political decision on the required end state, a clear prioritized list of essential functionalities to be established, and above all an integrated approach through coordination, which has to start well before the IC addresses conflict. This coordination should take into account the strengths and weaknesses of all parties involved [IOs, governmental organizations (GOs), major NGOs, and the military) as well as an open mind for possible solutions, creativity, and lessons learned from the previous operations. The key to effective coordination lies in joint civil military mission planning. One common planning platform for the civil military cooperation (CIMIC) should be intelligence activity, which could provide shareable and accessible databases for involved civil and military components in the area of a complex emergency.

Since 9/11, many nations have already addressed a need for closer transnational cooperation; national police forces work closer together in the international arena, and given time, it should be possible to enhance intelligence sharing in CEOs to reach the overall goals more efficiently and, ultimately, the desired end state.

Despite the global dimensions of contemporary security, the national (state) dimension remains a key factor: The state ensures security to its own citizens with an active national security system. The effectiveness of this state is measured not only by the level of protection of its own fundamental societal values from external and internal threats, but also by the ability to provide economic, political, scientific, technological-technical, social, cultural, ecological, and other well-being issues for the population. Above all, the effectiveness of the state is measured by its ability to encourage sustainable development. The fundamental structural elements of a national security system are the operational capabilities of the society that can provide its own security. Contemporary political science has recognized the municipality/province as the main local governance entity, which is needed to enable the overall societal security for its own population. Municipality/province capability for providing comprehensive security is inversely proportioned to its vulnerability to assess and manage security risks.

Holistic societal analysis requires the proper selection of demographic, social, political, economic, and environmental variables at the municipality/province level to measure the potential of each security dimension and local capability for societal reconstruction as a whole.

A systemic approach toward a complex emergency situation has enabled the IC to recognize the crisis area, gain and understand root causes of the conflict, as well as implement the decision-making process for intervention and planning for execution of the operation in a timely manner. The effort to build an integrated approach to CEOs requires a high-level approach among all parties involved. An integrated approach can be facilitated by the establishment of permanent representation in each other's organizations without losing its own identities and values and by respecting those of others. A better understanding between security organizations is needed, which means each other's strengths and weaknesses, before a conflict develops and materializes. Recognizing the different fields of expertise and coordinating an integrated approach to a complex emergency by permanent representation within each other's institutions will most certainly be a major step forward in addressing complex emergencies. The ever more complex crises demand a reconciled and timely activity of security instruments, where the exercises offer ample possibilities for ensuring suitably trained components of peacekeeping forces for the efficient implementation of peace operations.

1.2 EXERCISES

People want to be aware of and prepared for the incoming threats existence, families, tribes, religious groups, districts, municipalities, provinces, as well as whole countries. It has always been a question about their perception of security and readiness to react properly to protect themselves as well as their society and societal values. The capability of each society could be described by the ability to protect a certain level of the main security dimensions, which are as follows: demography, economy, social, environmental, and political dimensions. A balanced interaction between these dimensions ensures desirable and predictive societal development. In the case of unexpected, unscheduled, unplanned, unprecedented, and definitely unpleasant events [CRI01], society has to have appropriate countermeasures in place to neutralize undesirable effects. Forces that must apply proper activities for protection of the population and societal values represent the operational elements of the national security system in each country worldwide. Operational elements consist of military, police, and civilian forces. All of them have skills such as decision making, communication, situational awareness, team work, and stress management. Maintenance of these skills requires constant exercise with different training methods, such as live exercise, command-post exercise (CPX), crises-management exercise (CMX), and computer-assisted exercise (CAX).

1.2.1 Military and Civilian Exercises

Society consists of the following structural elements: civil society, military forces, and civilian elements of the state national security system. These elements perform decision-making functions, operational functions, intelligence functions, advisory functions, and administrative functions. They all need to have proper education and training to provide timely and appropriate responses to the overall threats to the main societal dimensions.

Military education is defined as the permanent process of examining the staff, units, and force components with regard to their capabilities to conduct a given mission or task effectively and efficiently. It also investigates the validity and reliability of systems, procedures, programs, and objectives. It includes analysis, assessment, feedback, and lessons learned. Military exercises prepare commands and forces for operations in peace, crisis, and conflict. Therefore, the aims and objectives of military exercises must mirror current and anticipated operational requirements and priorities [NOR06].

The civilian side of the security system has a similar approach with different training objectives, different means, different procedures, and different resources. The most valuable methodology for all security elements is to have a common exercise with the scenario, which can provide realistic training conditions for the acceptable and achievable exercise objectives. A common understanding of possible security threats, outcomes, and overall procedures involves increasing the situational awareness among all participating security elements and strengthening the ability of the national and global security systems to manage threats to modern society. States and security organizations plan and conduct exercises at strategic, operational, and tactical levels to achieve the following:

- Enhance operational capabilities, readiness, standardization, and effectiveness of command structure and force structure, as well as assigned forces
- Demonstrate capabilities and ensure the effective integration of assigned forces

- Enhance civil–military cooperation
- Enhance the production of operationally effective, interoperable, and capable forces for crisis response operations (CROs)
- Complement the internal training programs
- Support the evaluation process [NAT07]

1.2.2 Live Exercises

Live exercise is a training method, which is planned and conducted on a tactical level for individuals, teams, military units, and their staffs. This typical situational training involves real tools, weapons, and a real training area; here, collective training overlaps individual training. If the participants belong to a civil security element or units, then the normal training events are linked to a crisis situation as follows: search and rescue, fires, earthquakes, floods, spread of diseases, tsunami, and big traffic accidents. All of these events happened already, so different teams have an opportunity to learn, train, and evaluate their knowledge and skills. Similar events occur with military units, which prepare themselves for engagements in combat missions and for peace-support activities. Combat missions primarily deal with the tactics and procedures to defeat the adversaries; peace-support activities are more connected to tasks of maintaining peace by patrolling, establishing check points, protecting convoys, helping the affected population, and giving military assistance to the host nation. Because real weaponry systems and equipment are used, the planners of military exercises need to be aware of environmental risk management.

1.2.3 Command Post Exercises

CPXs have been designed primarily for training of headquarters (HQ) as follows:

- Efficient execution of operational missions and tasks
- Development of knowledge and procedures, which will allow for training the audience to participate in real operations
- Handling with computer and information systems
- Evaluation and improvement of operational procedures

Most CPXs are dedicated to military commands. For example, the North Atlantic Treaty Organization (NATO) has been using a system approach to training (SAT) model [NAT07], which was designed to provide a more effective, efficient, and economical approach to training by focusing on mission-essential tasks, their respective requirements, and necessary training objectives. The SAT model encompasses four major steps: analysis, design, conduct, and evaluation (Figure 1.2).

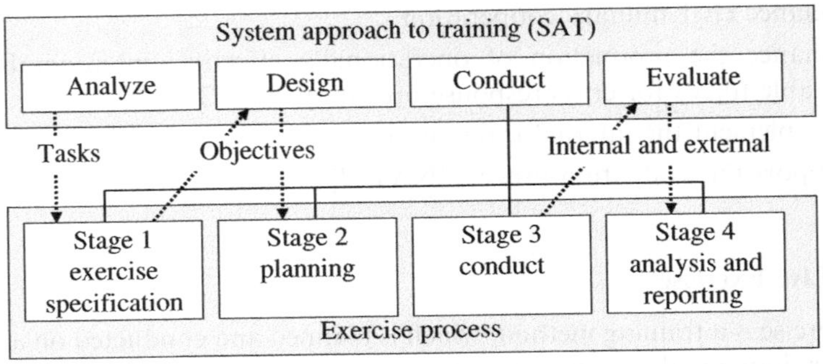

Figure 1.2: SAT model steps and inter-relationships among exercise process stages.

This first SAT step for the commander and staff is to analyze the missions and essential tasks as well as, the necessary capability requirements to achieve them. The Design step should ensure the required training objectives are met. Training design should address, inter alia: objectives to achieve; contents to deliver; methods to use; audience(s) to consider; resources required; establishment of schedules; requisite evaluation; and determination of costs. The Exercise Process Operational Conduct Stage corresponds to the SAT Conduct Step and includes the four phases of the NATO Exercise Training Model as follows: individual and collective training, crisis response planning, execution, and assessment. The first phase is preparing and training the audience, staff, and command group, in terms of theoretical preparation for planning phase. Crisis response planning is a process in which create staff create, with the commander's direction and guidance, a mission analysis and course of action proposal, a desirable concept of operation, and an operational plan for the execution. The execution of the operation is driven by injection of different incidents in the decision-making process. Sometimes, vignettes have been used for these purposes. The CPX process could be used on a tactical or operational level, with pure military participation or in a civil-police-military training environment. The last phase in the process is the assessment of the execution, where designated evaluators assess the quality of the training audience performance in accordance with overall accepted criteria. CPX could be executed on the real training areas with a combination of the live exercise, in laboratory conditions using only military maps or geographical information system (GIS) tools, or with the help of a simulation model on the CAX.

1.2.4 Computer-Assisted Exercises

The execution of training, which demands the movement of bigger units, is more expensive not only because of transportation costs but also because of costs for military activities on the training area, consumption of fuel, and maintenance of the training equipment. The need for response forces is often given on short notice, which is a critical factor of a quick response. This factor is especially important for the countries that are sending forces to areas with complex emergencies. Military, police, and civil components can use the same information technology (IT) and communication techniques to decrease training costs and support unity of preparation for the peace operation. All these reasons contribute to the decision for using simulation models to support CPX. Whenever computer simulation modeling is included in the exercise process, planners are conducting CAX, which does not deal with environmental risk management. The main training area is a synthetic environment where all planned units are displayed by a simulation model.

A CAX can be defined as a type of synthetic exercise (SYNEX) where forces are generated, moved, and managed in a simulation environment based on the commands from the exercise participants. Therefore, CAX support is often thought to be limited to installing and running a military constructive simulation during a CPX. In this example, CAX support is used to replace or to help response cells, higher level commands (HICON), and lower level commands (LOCON) to evaluate the possible outcomes of the decisions or requests coming from the training audience (TA) by running a set of stochastic processes. However, CAX is in essence a CPX where electronic means are used as follows:

- To immerse the TA in an environment as realistic as possible
- To help the exercise planning group (EPG) and the exercise control (EXCON) staff to control the exercise process (EP) so that it achieves the objectives as effectively as possible

Therefore, the definition of CAX should not be limited only to the usage of modeling and simulation. CAX ensures the high quality of individual and collective training on decision-making processes on tactical, operational, and strategic levels.

The application of simulation models on CAX represents an educational method, which dynamically introduces the operational conditions of real systems in a synthetic environment. A dynamic training system consists of digital terrain and equipment that allows the exercise participants to gain new knowledge, skills, and behavior. Each CAX is also a research method, because it introduces and instructs the following lessons:

- Novelty of the problem
- Importance and applicability of solutions for the practice
- Level of interest in problem-solving processes
- Available equipment and other research conditions
- Actuality of research results
- Possibility to find solutions for the decision-making problems by research

Through the process of CAX, we are undoubtedly optimizing current staff procedures and decision-making processes in synchronization with all other stakeholders in the area of responsibility.

1.3 MILITARY SIMULATION

Simulation justifies itself, perhaps the best for military, because it reduces the cost of training considerably. Simulation is also the only way to test and train for some special environments, such as nuclear events, biological and chemical contamination, and operations that require large-scale mobilization and movement. Creating and maintaining such environments is simply impractical in real life without interfering with natural life. These hostile environments also imply casualty risks even within training. Therefore, simulation is not only a cost-effective approach for military training but also is the only practical and safe way for testing and training in special and hazardous environments. Therefore, military simulation systems have been used extensively. They can be categorized into three broad classes as shown in Table 1.1.

- Live simulation refers to a simulation that involves real people in real systems. For example, two pilots can be trained for dog fighting by using real aircrafts in the air. In this case, the aircrafts and the pilots are real, but the interactions between the aircrafts are simulated, and the simulation decides how effective the pilots and the aircrafts are against each other. Similarly, all the weapon systems can be equipped with emitters, and all the equipment and personnel can be equipped with sensors. If the weapons are aimed and fired correctly, the emission by the emitters can be sensed by the sensors, which indicates a hit and a kill based on some stochastic processes. By using such a technology, troops can exercise in a very realistic environment. Both of these are examples for live simulation. Civilian applications of live simulations are also available. For example, fire fighters, personnel in oil rigs, nuclear power plants, and cruise vessels can also be trained with live simulation systems.
- Virtual simulation refers to a simulation that involves real people in simulated systems. The best examples for these are aircraft and tank simulators, where a simulator and not a real system is used to train a pilot or tank crew. Airplane simulators are also widely used to train airline

Table 1.1: Military simulations.

Category	People	Systems
Live	Real	Real
Virtual	Real	Simulated
Constructive	Simulated	Simulated

pilots cost effectively without risking the life of pilots and expensive airplanes.

- Constructive simulation refers to a simulation that involves simulated people operating in simulated systems. Combat models that compute the possible outcomes of the decisions made by the headquarters fall into this category. In these simulations, people and the unit are also simulated, as well as combat systems and the environment. Constructive simulation is also used for civilian purposes. Crises-response operations, disaster-relief operations, economical situations, and communications systems can be simulated by using constructive simulations.

Although the interaction of these categories of simulation systems is possible by using the state-of-the-art technology and all of them (i.e., live, virtual and constructive simulations) may be used during the execution of a CAX, a military constructive simulation system constitutes the core of a CAX. The definition of constructive simulation states that the people operating the simulated systems are also simulated. However, real people enter the commands to stimulate these simulations. Constructive simulations are designed to find out the possible outcomes of the courses of action taken by the real people. They are constructed by many models, and often stochastic processes calculate the results of interactions between the entities or units in a theater. Constructive simulation systems can be classified into two categories according to their resolution as summarized in Table 1.2.

- *High-resolution simulations are entity-level simulations* in which singular military objects (e.g., a soldier, a tank, or an aircraft) are the primary objects represented. They are designed for the lower military echelons, such as platoon, company, and battalion. The resolution of terrain data is higher sometimes up to the plans of individual buildings. However, the simulated terrain is often limited to 200 km × 200 km.
- Highly aggregated simulations are aggregate-level simulations in which collections of military assets (i.e., units) are the primary objects represented.

Table 1.2: Military constructive simulations.

Category	Level	Objects	Echelon	Terrain
High resolution	Entity	Singular objects (e.g., tank and troop)	Tactical (e.g., platoon and company)	High resolution, 200 km × 200 km
Highly aggregated	Aggregate	Units (e.g., battalion and company)	Operational (e.g. corps)	Low resolution, 4000 km × 4000 km

They are designed for the higher military echelons, such as corps level. They use lower resolution terrain data, but they can simulate areas as large as continents.

The gap between high resolution and highly aggregated simulation systems is closing. State-of-the-art entity-level simulations can be used up to operational levels (i.e., corps) and can simulate regions as large as 2500 km × 2500 km. On the other hand, aggregate-level simulations tend to simulate entities such as a single troop and a tank. However, the nature of CPXs at operational and tactical levels forces us to assess the situation at aggregate and entity levels, respectively. Therefore, the current trend to have either a single system that can aggregate and de-aggregate the units and entities or a multiresolution federation of entity- and aggregate-level simulations.

The constructive simulation systems can also be categorized based on their functionalities as follows:

- Service models are the simulation systems developed for the needs of a single service (i.e., Army, Navy, or Air Force).
- Joint models are either the simulation systems that fulfill the requirements of all services or the federations that are made up of service models.
- Expert models are developed specifically to simulate certain functionalities in the theater, such as logistics, intelligence, electronic warfare, homeland security, and space operations. They can also join federations.

Global security complexity, security components of the IC, and computer simulation modeling represent the main ingredients for the CAX. The motivation for putting different content together was to share and help the others improve the quality of training and to influence future response of IC on security challenges around the world.

1.4 SCOPE OF THE BOOK

In this book, the authors elaborate on the operational, technical, and organizational aspects of a CAX. The three pillars of expertise for preparing and conducting a CAX are as follows:

- People from an operational background: Either they can be responsible for planning and/or control of exercises, or they can be the training audience in a CAX. They can also be decision makers that use modeling and simulation as decision-support tools for operational or acquisition-related decisions. Combat system designers who use combat models to evaluate the system performance can also fall in this category.

- CAX support personnel: These people are from both a technical and an operational background and are responsible to facilitate CAX support tools for exercises or experiments.
- CAX tool developers: These people are from an operational, operational research, computer science, networking, or communications background. They analyze the requirements, design the models/systems and the other CAX tools, as well as implement, validate, test, and verify them. We count researchers who perform academic research related to this field in this category.

This book is designed to be both a textbook and a reference book for everyone that has a role in CAXs. Therefore, we first examine the contemporary security environment to provide a better insight about the phenomenon modeled by military simulation systems. We also provide detailed information about CAX-related statistics, probability theory, and modeling. This information is essential not only for the system designers and developers but also for the users because a simulation is the imitation of a real environment and systems by using many simplifying and generalizing assumptions. A user should understand the implications of those assumptions; otherwise, he/she may get a misleading perspective after computer-assisted exercises and experiments. We also explain the details about organizing a CAX as well as all the CAX support tools, which include but are not limited to modeling and simulation. This information is required by CAX planners and exercise control staff, as well by the industry that will provide tools to support these organizations and exercise processes. The bottom line is that we aim to provide a complete view of all the aspects related to a CAX. Therefore, we believe this book is comprehensive.

1.5 STRUCTURE OF THE BOOK

This book is designed as a self-contained textbook for both developers and practitioners. Therefore, it has two parts. First it introduces the fundamentals and key issues related to CAXs. We write this part such that it can be understood by both people either from an operational or a technical background. We use examples from the military domain when explaining scientific issues. That technique can help the technical issues to be understood better by the operational personnel, and the operational requirements can be understood better by the developers. In the second part, we present advanced information related specifically to CAXs. We also mention the gaps in fulfilling the requirements. We believe that a 100% solution will never exist in this field. As new systems are developed, there will also be new requirements. Therefore, we will always have some gaps in satisfying users. Actually, we believe that the gap has been widening for the last decade, although the international community has been putting forth tremendous effort to develop new tools and systems.

The contemporary security threats are changing, and nonkinetic warfare is gaining more and more importance. In the second part of the book, we examine the gaps and make some predictions for the future.

The first part of the book consists of six chapters, which include the Introduction. The second chapter explains warfare and how it has evolved. That is an important part of this book because the aim of a CAX is to train headquarters to operate in a war. Warfare has changed dramatically for the last decade. A new generation of warfare that requires a comprehensive approach and the cooperation of both military and civilian organizations is introduced. We explain these approaches and what they mean for CAXs in Chapter 2.

The fundamentals about probability theory and statistics are explained in Chapter 3, and prerequisite knowledge for combat modeling is provided. This chapter is important not only for technical but also for operational staff involved in a CAX. We elaborate how and why the results of a simulation run can change from another run, why selecting a correct statistical distribution is important, and how a better analysis of the results can be performed.

Chapters 4 and 5 discuss simulation and distributed simulation, respectively. Simulation tools are a key category of CAX tools. We explain what simulation means and various approaches in designing a simulation. Distributed simulation is an approach to build more effective, practical, and higher fidelity simulation environments. It has gained importance during the last decade. We also explain the history and components of distributed simulations.

Chapter 6 is the final chapter of Part 1; this chapter explains experimentation. CAX support tools can be used for experimentation and analysis. For example, operational plans can be analyzed before an operation or the performance of alternative combat systems can be evaluated before an acquisition by using military simulations. Experimentation and analysis need a structured approach and a good understanding of the simulation system. We elaborate on key issues for the correct execution of an experiment by using CAX support tools.

The second part of the book consists of five chapters and starts with an explanation of the CAX architecture. Perhaps the most challenging part of a CAX is the design of a CAX structure and the creation/management of the organizational entities in a CAX structure, which is typically made up of three main components: training audience, exercise control, and exercise support. The composition of these entities changes from a CAX to another, and the successful conduct of a CAX depends more on the correct composition of exercise components than on the efficient tackling of technical issues. Chapter 7 provides generic exercise constructs.

Chapter 8 is about a CAX process. A CAX process typically consists of specification, planning, execution, and action review phases. This generic approach is applied by many nations and international organizations when planning and conducting an exercise. We elaborate on this method and highlight key issues related to a CAX process.

The challenges of combat modeling are described in Chapter 9. Both traditional and contemporary schemes are used to model the kinetic aspects and environment in a war. Also described areas the difficulty and challenges related to modeling generation warfare.

Military constructive simulation systems are categorized, and the ones best known to us are introduced by short summaries of the systems in Chapter 10. Vendor companies or developers of the systems provide these short descriptions about the selected military constructive simulation systems. Please note that the list in this book is not an exhaustive list, but it provides examples from all over the world. A CAX process is supported not only by combat models but also by other CAX support tools, like CAX planning and management tools. Those tools are categorized, and some important examples are introduced in Chapter 10.

Technical challenges as well as communications and information system (CIS) requirements are the main hurdles to pass in a CAX process. Bandwidth and computational power are generally scarce resources that need to be used efficiently. Fault tolerance is also a requirement. The execution phase is typically long enough to experience several CIS and/or CAX support tool failures. The CAX and CIS systems must be fault tolerant to keep these failures transparent to the EXCON staff and training audience. Technical challenges, risks, and risk-mitigation techniques are explained in Chapter 11.

Chapter 12 describes the organization and management of the EXCEN, where an efficient and effective EXCEN is essential for the success of CAX.

1.6 ELECTRONIC RESOURCES FOR THE BOOK

The book has a website available at: http://www.caxbook.cayirci.net/. At this site, you can access the following:

- Slides: Microsoft PowerPoint slides prepared for the content of the book are available at this link. The slides are organized for a 12-week course where the book provides the text. They are kept up to date.
- Updates table: The updates and corrections for the book are listed in this table. If you have any comments to improve our book, please e-mail them to us by using the "contact us" link at the book website.
- The interest group: This link provides you with a user name and password to access the interest group for the content of the book.
- Useful links: The links for useful websites are provided at this page.

1.7 REVIEW QUESTIONS

1.1 What is an exercise?

1.2 What are the differences between a live and a command post exercise? What are the advantages and disadvantages of each of them?

1.3 What are the steps of the systems approach for training? Discuss each step. What are the tasks related to each of these steps?

1.4 How can you categorize military simulation systems?

1.5 Is live or virtual simulation more useful for civilian training? What about military training?

1.6 Describe a scenario where constructive simulation systems can be used for civilian purposes.

1.7 How can you categorize constructive simulation systems? Discuss each category. Is there another way to categorize constructive simulation systems?

1.8 If high-resolution constructive simulation systems provide a better fidelity and current hardware capabilities can fulfill the computational requirements of high-resolution systems for any scenario, do we still need highly aggregated systems? Why?

2

CONFLICT AND WARFARE

The main tool for providing training conditions on a computer–assisted exercise (CAX), as described previously, is a realistic scenario, which is a description of fictitious conflict developments and all involved parties. The scenario is enables data for computer modeling. Thus, the main focus of the exercise scenario is artificial conflict or unstable interactions among societal security dimensions of a synthetic environment, which leads to war.

Conflict is a natural consequence of the clashing of interests (positional differences) over national values of some duration and magnitude (independence, self-determination, borders and territory, and access to or sharing of internal or international power); conflict occurs in a certain time and space between at least two parties (states, groups of states, organizations, or organized groups), which are determined to pursue their interests and achieve their goals. At least one side is an organized state [HII07].

A conflict is a situation that often involves violence or the threat of violence. With regard to the degree and frequency of violence, the spectrum of conflict ranges from low intensity, to medium intensity, to high intensity. Concerning the type of hostility, nonviolent conflicts, nonconventional and conventional, as well as nuclear hostilities are distinguished. Low-intensity and medium-intensity conflicts are limited political and military interventions aiming to achieve political, economic, social, and psychological objectives always in a geographically limited area and with limited employment of armed forces. War fighting is not the main instrument to pursue strategic goals. War is a high-intensity conflict. In the traditional sense, war is an aggravated social conflict in which social groups, movements, organizations, states or coalitions of states

Computer Assisted Exercises and Training: A Reference Guide, by Erdal Çayirci and Dusan Marincic

wage fighting in an organized manner to accomplish their fundamental objectives through a prevailing employment of armed forces. It is a complex situation in which political, economic, psychological, and other forms and contents of confrontation take place simultaneously to war fighting, of which some confrontations cannot be kept under control. War is the continuation of politics with other more violent means, and it is the most violent form of promoting political interests. With regard, to the area, world, regional, and local wars are distinguished.

The general characteristic of contemporary wars is the integration of most military and nonmilitary resources of warring factions. War is often asymmetric. States can enter conflicts, as can noninstitutional organizations and individuals. On the one hand, there are states and formal armed forces, and on the other hand, there are various organizations, such as terrorist organizations. In some situations, civilians attack civilians and civilian infrastructure.

Contemporary wars are characterized by high casualty rates of one party or both parties involved in a conflict, as well as by collateral damage. The initial stage of war is shorter, because of the employment of rapid response forces that can be deployed quickly to distant areas, and because of the possibility of air raids from a large distance against military forces, communication, traffic infrastructure, information centers, and other targets [DOC06].

Carley and Christie [CAR92] have mentioned that in modern political systems, especially in democratic ones, permanent tensions exist among "top-down" forces and "bottom-up" tendencies. These tensions comprise forces of centralization and decentralization. Trends toward *centralization* inside national borders have been linked the following:

- Ascension of modern state
- Increased control by the central government on long-term policy and allocation of sources on territorial basis
- Cultural homogenization
- Additional administrative control over occupations and bureaucracy

Trends toward *decentralization* are often the result of the following:

- Political agreements on ideological regionalism or ethnic nationalism
- Measures for promotion of the regional economical development
- Measures for mobilization of local sources and consecutive obligations
- Functional responses on government overload, bureaucratic indifferences, or often mistakes of centralized political initiatives

After 1989, permanent tensions such as previously mentioned created conditions for disintegration of the Soviet Union, Yugoslavia, and Czechoslovakia in more or less violent way. The international community (IC) has also witnessed tensions in India (Punjab, Kashmir, Assam) and in Sri Lanka, are very persistent where the Tamils and in East Timor.

Vlassenroot [VLA95] has described, on the basis of the origins of conflicts, the following types of conflicts, about which it needs to be stated that in practice conflicts are made up of a combination of the following types:

- **Legitimacy conflicts**: These conflicts develop from a regime's lack of legitimacy, which leads, for example, to an absence in political participation and a problematic distribution of wealth and welfare. When participation in power is not possible through the system itself, it may be felt that the only option is to attempt a coup or to challenge the regime with violence. Often, wars between regimes result from this type of conflict.
- **Transition conflicts**: Characteristic of a change in the system, which can be initiated by the power center itself or enforced by opposition forces or movements, are intensified struggles among rival actors with differing interests. In this situation, the power to force your own views of rule and policy on the other actors is at stake. When the transitional process does not produce the outcome that was hoped for, there will still be a chance that more conflict will follow.
- **Identity conflicts**: These conflicts are a consequence of searching for one's own identity, protecting one's own safety, or the lack of access to political power and economic sources. At their root are ethnic, religious, tribal, and linguistic differences. What are termed ethnic conflicts are often conflicts among various elites who recognize a legitimacy for their battle for power in ethnic or nationalistic factors and can use it to mobilize the population.
- **Development conflicts**: These conflicts are a consequence of the growing gulf between the rich and the poor, or they can be a result of widespread impoverishment of deprived groups and sectors of society and regions.

Enloe [ENL90] suggests that each conflict, especially ethnic conflicts, is complex. Conflict could be influenced by culture, institutional structure, division of classes, and external factors. Ethnical contrasts exist because of distrust among cultures and a distinction in power. They are the result of a misuse of power and a lack of force. Improper use can lead to an unjust share of state power and public services. Insufficiency leads to an escalation of violence without increased perception of citizens that they are secure. A real solution for each ethnical conflict must involve a new evaluation and reorganization of military forces and police forces. Sustainable conflict resolution demands the redistribution of security and influences in the country. When the culmination of the conflict cannot be solved by peaceful means, war will occur.

2.1 PARADIGMS OF WAR

The essence of war remains the same no matter how one defines that essence. Carl von Clausewitz's definition of war as "an act of force to compel our enemy to do our will" is as valid today as it was 200 years ago. Violence is what turns a conflict into a war [HOW94].

Gattuso [GAT06] has discussed two broad theories of war: the *theory of attrition*, which advocates material superiority, mass, and attacks on the enemy's strength; and *maneuver theory*, which attacks the enemy's will, critical weakness, and cohesion.

The theory of attrition is essentially concerned with the destruction of the enemy's mass and physical forces. It searches for the enemy's strength and center of gravity. The attritionist seeks victory by attempting to destroy the forces in the field, which necessitates a focus on battle the tactical event wherein those forces are engaged and destroyed. Doctrine, force structure, and personnel are accordingly written, procured, and trained toward the decisive battle where the enemy is brought to the field, met cleanly, and defeated decisively. Battle is the preferred method for winning wars. The key concepts in attrition warfare are those of initial-force ratios, which involve the real or perceived numerical and material superiority of one side or the other; loss ratios, which include the rate of losses in men and material by both sides as a result of battle; and fractional exchange ratios, which are expressed algebraically as the loss ratio over the initial-force ratio. Attrition warfare seeks to improve the force ratio by achieving and sustaining an acceptable loss ratio over the enemy.

The characteristics of attrition theory include an emphasis on the superiority of competing forces, a focus on technology and equipment, primary attention by all command levels to the tactical level of warfare, and the destruction of the enemy's forces by impact and superior firepower. Because attrition theory focuses on force relationships and relative measures of technological advance, attritional military organization views warfare as scientific, measurable, and definable. The focus is on the quantifiable and tangible elements of war. Warfare is approached systematically.

In contrast to attrition theory, which targets the enemy's physical forces, maneuver theory concentrates on outperforming the enemy's thought processes with the intent to destroy force cohesion. The enemy's mental, moral, and physical stability is the object of maneuver theory; its focus is on the enemy's ability to observe, orient, decide, and act (OODA). This ability may or may not entail a primary concern with the enemy's forces in the field. The maneuver theorist eyes the enemy closely and adopts whatever methodology works to preempt, dislocate, or disrupt it. This style of war fighting carries enormous consequences for doctrine, force structure, personnel requirements, and leadership. The first characteristic of maneuver theory is a tendency toward decentralization. The primary need is to work quickly through the OODA loop; passing information up and down a centralized chain of command is inimical to deciding and acting faster than the enemy. Maneuver theory produces a military notable for generating and then thriving on confusion and disorder in enemy organizations.

Because decentralized command arrangements depend on a local subordinate unit initiative, to solve the situation at hand, the enemy is likely to discern no regular pattern of operations. The maneuvers military disdains standardized or traditional solutions to problems. A maneuvers military places a strong emphasis on the quality, trust, and independence of thought and action of and within its officer corps. Discarding a dependence on formulas or fixed solutions requires lower level leaders who can act individually based on the situation, personalities, and intentions involved. A great degree of trust is required from senior leadership. With the focus on the enemy's thought processes, as well as the requirement for high-initiative, creative, innovative, and trustworthy leadership, a maneuvers military tends to be "people centered" in contrast to an attritions military, which by nature tends to focus on technology and hardware.

Maneuver militaries ensure that their officers are sufficiently educated in the profession of war, look with a close and stringent eye to promotions and other reward systems, and place emphasis on rigorous historical study. As a corollary, they prefer less complicated technologies and weapons; technology is the trade, whereas people and leadership are the professions. When it succeeds, maneuver war fighting accomplishes its goal more decisively than does attrition. The collapse of the enemy, wrought through the destruction of mental, moral, or physical cohesiveness, is more dramatic. Panic, rout, or a resigned passivity are the hallmarks of an enemy defeated by maneuver warfare.

2.2 EVOLUTION OF WARFARE

Even though warfare may seem to change, it remains as Clausewitz defined it, just as the chameleon; whatever color it adopts, it remains the same animal still. All historians do agree, however, that a systemic change in the conduct, if not in the nature, of war was brought about during the 19th century by the technical transformations of the industrial age. So long as society depended on manpower, waterpower, wind power, and animal power for its energy sources, warfare had consisted basically of battles or sieges conducted by armies whose size was narrowly constricted by logistical limitations.

Technical innovations had indeed made incremental changes. The stirrup made cavalry a controllable instrument for organized battles as well as for sporadic raiding. Mobile heavy artillery transformed siege warfare, as it had been conducted from antiquity until the end of the Middle Ages. The combination of the ring bayonet with the flintlock provided a force multiplier for infantry, Which made every soldier his own musketeer as well as pike man. The same kind of force multiplying effect was gained when Jean Batiste de Gribeauval's French army reforms resulted in a new generation of more mobile and accurate field guns in the mid-18th century.

Throughout the agrarian age war consisted of battle and the search for battle. And battle consisted, or was viewed to consist, in corps-a-corps fighting with "cold steel." All developments in firepower were perceived as ancillary to this. Artillery was developed to make it possible for infantry to close with the enemy, not to make it unnecessary. Infantry volley-fire was always preliminary to a charge. Around this perceived necessity for the decisive corps-a-corps encounter, a whole military culture developed. In this social hierarchy, those who delivered the "shock" in battle, such as the cavalry and elite infantry, were at the top. All ancillaries, which include artillery, took their places lower down the pecking order.

Napoleon Bonaparte became an artillery officer because he did not have the social standing to get into the infantry or cavalry. It is interesting to speculate whether he would have been so innovative in the conduct of war if his pedigree had been good enough for the cavalry. During the agrarian age, the only fundamental changes that occurred in the conduct of war were the results of social and political factors rather than technological innovation. Furthermore, it was only the development of the bureaucratic state in early modern Europe that made possible the development of professional, disciplined, long-serving armed forces, especially navies. And it was the French Revolution, not any technological breakthrough, that made possible the Napoleonic campaigns, which not only introduced a new operational concept into warfare but also overthrew an entire political order in Europe and prompted Clausewitz to foresee a new era of "absolute" war.

Howard [HOW94] has emphasized that Clausewitz prophesied better than he knew. The industrial age of which he was totally unaware did indeed transform warfare. It did so by increasing the range, accuracy, and lethality of weapons, whereas logistical developments, in particular the railroad and the telegraph, made possible total war on a scale such as even Clausewitz had never conceived. Between them, these two developments produced the destructive deadlock of World War I. The development of firearms, magazine-fed repeating rifles for the infantry, and breech-loaded guns for the artillery made it impossible for armies to come close enough to obtain the kind of decision that battles had always been fought to achieve. The development of railroads, telegraphs, and supply systems meant that the size of armies was such that their limitations were determined only by the size of the population and the economic capacity of the state to train and sustain them. The result was what might be called Total War, where the entire resources of the state were mobilized to sustain armies in the field whose only formula for victory was attrition and whose commanders were military managers. The object of operations became, not the destruction of the enemy army on the battlefield, but, by engaging that army in prolonged and inescapable battles of attrition, to bleed the opposing society to death.

2.2.1 First, Second and Third Generation of Warfare

Toffler explains[1] that "[S]ociety needs people who take care of the elderly and who know how to be compassionate and honest. Society needs people who

[1] Alvin Toffler interviewed by Norman Swan, Australian Broadcasting Corporation Radio National, "Life Matters," March 5, 1998.

work in hospitals. Society needs all kinds of skill that are not just cognitive; they're emotional, they're affection. You can't run the society on data and computers alone." In his book *The Third Wave*, Toffler describes three types of societies, which are based on the concept of "waves" Each wave pushes the older societies and cultures aside. The first wave is the society after agrarian revolution and replaced the first hunter-gatherer cultures. The second wave is the society during the Industrial Revolution (late 1600s through the mid-1900s). The main components of the second wave society are nuclear family, factory-type education system, and the corporation. Toffler writes: "The Second Wave Society is industrial and based on mass production, mass distribution, mass consumption, mass education, mass media, mass recreation, mass entertainment, and weapons of mass destruction. You combine those things with standardization, centralization, concentration, and synchronization, and you wind up with a style of organization we call bureaucracy." The third wave is the post-industrial society. Toffler would also add that since the late 1950s, most countries are moving away from a second wave society into what he would call a third wave society. He coined lots of words to describe it and mentions names invented by him (super-industrial society) and other people (like the Information Age, Space Age, Electronic Era, Global Village, technetronic age, scientific-technological revolution), which to various degrees predicted demassification, diversity, knowledge-based production, and the acceleration of change (one of Toffler's key maxims is "change is non-linear and can go backwards, forwards and sideways").

2.2.2 Fourth and Fifth Generation of Warfare

Henry and Peartree [HEN98] have written that whereas the Tofflers' theses are less than perfect, they are generally correct with respect to the goals of warfare imposed by the prevailing socioeconomic frameworks of the various epochs. Successful preindustrial war was generally predicated on the seizure of territorial assets, control of them, or both. Successful industrial age war was about reducing the means of production and out-manufacturing one's opponent. If the analogy holds, the advance guard of theorists and defense analysts contend that future wars will be waged for control of data, information, and knowledge assets. Weapons of war also reflect the dominant aspects of each era's socioeconomic paradigm. Rifled arms, iron-clad ships, machine guns, tanks, and aircraft depict the evolution of industrial age war. The precision-guided munitions, which are popularly known as the "smart bomb," herald for some weaponry of the information age. The deeper expression of any age, however, can generally be found in the organization and culture of the war-fighting community. Some propose that hierarchical command structures and ponderous military-industrial bureaucracies, which created to fit industrial were age needs, must now give way to the decentralized, "flattened" business network of the information age. But liabilities are associated with moving too rapidly to reengineer the force around new technologies without first considering interests and risks. The appearance of new weapons and new technologies has sometimes caused military leaders and theorists to make errors in judgment, misreading the meaning of the new technology and producing poor returns on the investment, whether on the battlefield or in the view of history.

Current theorists therefore conclude that the new mode of warfare ushered in by the information revolution will have sweeping effects on the conduct of war in the near future. Precision weapons will be directed at the enemy's decisive point(s) at the critical moment through "information superiority." Superiority, in turn, will occur through space, near-space, and ground-based sensing technologies that will transmit attack instructions in real time via a "system of systems" that links all parts of the battle space. Some even predict that the new technologies will penetrate, if not lift, the fog of war. The more radical of the theorists predict that information warfare will not only provide dominant awareness of the battle space, but also it will allow us to manipulate, exploit, or disable enemy information systems electronically. The intent here evidently is to knock an enemy senseless and leave him at the mercy not only of conventional kinetic also attack but of psychological operations aimed at controlling his perceptions and decision-making abilities during operations.

One of the most pictorial comparisons of different generations of warfare was made by Hammes [HAM07]. He stated that on changes in the political, economic, social, and technical fields, first-generation warfare culminated in the massed-manpower armies of the Napoleonic era. In the same way, the second generation used the evolution to an industrial society to make firepower the dominant form of war. Next, the third generation of warfare took advantage of the political, economic, and social shifts from an industrial to a mechanical era to make mechanized warfare dominant. Fourth-generation warfare uses all the shifts from a mechanical to an information/electronic society to maximize the power of insurgency. Fifth-generation warfare will result from the continued shift of political and social loyalties to causes rather than nations. It will be marked by a nets-and-jets war: Networks will distribute the key information, provide a source for the necessary equipment and material, and constitute a field from which to recruit volunteers; the jets will provide for worldwide, inexpensive, effective dissemination of the weapons.

The above-mentioned comparative analysis has been created on the basis of shifts of societal security dimensions. This statement proves the importance of permanent monitoring and assessment of the local potential in each societal dimension to understand causes and possible solutions for the conflicts in the area of the complex emergency. All the changes in security dimensions are providing a clear picture about threats to certain society, its vulnerabilities and capabilities, and most important security trends on the way to sustainable development. With the proper understanding of these relations, the IC can plan and manage, together with security instruments, any kind of crises response operations in crisis areas around the world.

2.3 OPERATIONS

Operations are a planned, organized, and well conducted activity by designated organizations, companies, political entities, or military forces to achieve desirable objectives in a specific time and space framework. This chapter will primarily focus on military operations. The effectiveness of armed forces in peace, crisis, or conflict depends on the ability of the forces provided to operate together effectively and efficiently.

Operations should be prepared for, planned, and conducted in a manner that makes the best use of the relative strengths and capabilities of the participating countries and the forces they offer for the operation. A common doctrine supported by standardization of equipment and procedures, which is validated through participation in joint and multinational training exercises, provides the basis for the formations and units of a joint and multinational force to be able to work together. At the operational level, emphasis must be placed on the integration of the military forces and the synergy that can be attained. This synergy will have a significant effect on the ability of a joint force to achieve the commander's objectives.

Common sense and a balanced judgment are indispensable qualities for a military commander, but alone they will rarely ensure success in armed conflict. In most military operations, time is critical and information may be scarce and unreliable. The effects of danger and fatigue usually have an adverse influence on judgment, and unforeseen circumstances frequently upset the best-laid plans. To meet these exacting conditions, the commander's judgment must be backed by a sound knowledge of the doctrine as well as the advice and expertise of his staff.

The strategic level of war is the level of war at which a nation or group of nations determines national or multinational security objectives and deploys national, including military, resources to achieve them. The operational level of war is the level of war at which campaigns and major operations are planned, conducted, and sustained to accomplish strategic objectives within theatres or areas of operations. At the operational level, armed forces are deployed and employed to achieve military strategic objectives within a designated joint operations area (JOA). Normally, this action would imply sustained operations with simultaneous and/or sequential actions by committed forces. To attain objectives, operations have to be planned in the context of the operational environment, which requires a thorough analysis of the military strategic situation in the JOA and the ability to conduct planning under severe time restraints if situations change. Arrangements for cooperation and coordination with civil agencies, institutions, or organizations are important and part of the responsibilities at this level.

Operations by joint forces are directed, planned, and executed at three levels. They are directed at the military-strategic level and are planned and executed at the operational and tactical levels. Actions are defined as military-strategic, operational, or tactical based on their effect or contribution to achieving the specified objectives.

At the military-strategic level, armed forces are deployed and employed within an overarching political framework and in a synchronized fashion with other initiatives (e.g., diplomatic or economic) to achieve the strategic objectives. At the operational level and within a designated JOA, armed forces are deployed and employed in accordance with the campaign strategy to achieve military strategic goals. Normally, this action would imply sustained operations with simultaneous and/or sequential actions by committed forces. It is at the operational level that tactical successes achieved in engagements and operations are combined to achieve strategic objectives.

At the tactical level, forces are employed to conduct military tasks and gain military objectives. Successful accomplishment of these objectives is designed to contribute to success at the operational and strategic levels. The distinction among the military strategic, operational, and tactical levels of joint operations will seldom be tidy. The key to delineation is that normally a strategic authority allocates objectives and resources while setting necessary limitations. At the operational level, the commander orders the activities of the assigned formations in pursuit of the plan of campaign. At the tactical level, commanders employ units for combat to achieve the military objectives of the campaign. However, in the current security environment, traditional responsibilities are blurred, and the different levels of operations merge because even a small tactical incident could be considered politically sensitive, and thus, its conduct is under strict supervision at the strategic level.

2.3.1 Conventional Operations

Most military operations are designed to wrest the initiative from the adversary. To maintain and exploit the initiative requires a sense of urgency and determination to outwit the adversary. To do this, the commander must consider the manner in which the end-state (desired sustainable societal conditions in the area of complex emergency) can be achieved. However, many key concepts should be considered both before and during the conduct of the campaign, as follows:

- **Synergy and leverage**: These concepts can be obtained by the imaginative combined exploitation of different resources. This aspect could include the concept of the synchronization of firepower and maneuver.
- **Simultaneity and depth**: The simultaneity of simultaneous actions and also extension in depth shape future conditions and can disrupt an adversary's decision cycle as well as his ability to execute operations.
- **Offense and defense**: Offense and defense can be alternated in time, space, and level to put continuous pressure on the opponent.

- **Time management**: The operational commander must be able to stand back from the detailed conduct of the campaign to take time to identify emerging trends, grasp new opportunities, and detect potential threats.
- **Termination**: The conditions needed to terminate the campaign on favorable terms must be kept under continuous review.

Joint operations on land may involve maritime, amphibious, land, and air forces; such operations may need to be coordinated with operations conducted by forces from different countries. This action includes planning, task-organizing, deploying, and employing combat and combat support forces, as well as coordination of logistics and combat service support. A realistic appreciation of national capabilities, limitations, and priorities is essential, and with coordinated planning that builds combat power rapidly, a synergistic effect can be realized that will contribute to ending military operations quickly with minimal loss of life, reduced resource depletion, and possibly a limitation of collateral damage.

Land force operations will normally include elements of both offense and defense. Land force commanders apply combat power simultaneously across the depth, breadth, and height of the area of operations. To conduct such operations, land force commanders require concentration of forces in some areas by achieving an economy of effort in others. During initial operations, land forces may be required to defend while force buildup occurs. Even in sustained offensive operations, selected elements of the land force may need to pause, defend, resupply, or reconstitute while other forces continue the attack. Commanders at all levels must possess the mental agility to make rapid transitions between offense and defense. Defensive operations are usually undertaken to defeat or deter a threat to provide the right circumstances for offensive action. They aim to break the adversary's attack and destroy opposing forces to prevent the adversary from achieving his aim. Throughout defensive operations, commanders at every level must fight with imagination, energy, and aggression to seize or create opportunities to surprise the adversary; the aim is to attack, destroy, disorganize, and delay the opposing forces, not just at the point of immediate contact but wherever they can be engaged throughout the area of operation.

Offense is the decisive form of armed conflict. Offensive operations are conducted to defeat the enemy by the imposition of one's will and the application of focused violence to achieve both operational and strategic objectives. Offensive operations are characterized by audacity, concentration of combat power, speed of maneuver, reallocation of the main effort, rapid exploitation of weakness, effective but nonconstraining control of subordinates, maintenance of momentum, and simultaneous operations throughout the area of operations. The maintenance and employment of reserves are critical to all land operations.

In broad terms, the joint force commander must address three key operational functions if he/she is to be successful. He/She must be capable of preparing and shaping the battle space, in all dimensions, in a manner most likely to enable him to achieve the military end-state[2] at minimum cost. Also, he/she must be able to conduct the operation successfully (may or may not entail combat), ideally on his/her own terms and at a time and place of his choosing, and he/she must preserve and protect the military power of his/her own forces until his/her mission is concluded—which may be well after termination. Much of this can only be achieved by drawing together the constituent elements and components of the force, aided by a range of operational capabilities. Few, if any, of these activities are discrete. Shaping activities will continue throughout the entire operation.

Battle space is generally considered to be the environment, factors, and conditions that must be understood to apply combat power successfully, protect the force, or complete the mission. This area includes the sea, land, air, and space environments; the included enemy and friendly forces; facilities; weather; terrain; the electromagnetic spectrum; and the information environment within the operational areas and areas of interest. The battle space is the multidimensional environment embracing the JOA. The battle space is joint, in the sense that all military activities will be orchestrated in the context of a joint operation plan. Increasingly, these military activities cannot be planned or conducted in isolation. Other elements are also impinging on and influencing the environment traditionally exploited by maritime, land, and air components.

The need to integrate the efforts of the joint force with those of other agencies (for example, governmental and nongovernmental agencies and aid agencies) implies a greater convergence of the military, civil, humanitarian, diplomatic, and economic efforts at all levels. The effect, benign or otherwise, of the needs of the indigenous populations is also likely to be inextricably linked to the mission itself. Finally, the activities of the joint force will be under constant media scrutiny; indeed, in many cases, both the media and a range of civilian agencies will be there before the joint force has even deployed.

The boundaries of the area in which operations are conducted are becoming increasingly vague and blurred. At the higher, strategic level, it will be vital that the wider operational environment has been prepared and shaped in a way that will support the conduct of operations. Primarily, the legitimacy of intended actions and the general support of the international community must be self-evident to all. At the operational level, the joint force command (JFC) will seek to prepare and shape the battle space in the JOA by undermining the adversary's will and attacking its cohesion. The JFC will aim to erode the adversary's resolve, persuading him or her that military action is unlikely to be successful. To achieve this, the JFC staff members will draw on own intelligence assets to identify the adversary's vulnerabilities and weaknesses, and then attack them.

[2] This end-state is not a single point in time but is a transitional process to a new operational phase.

Interdiction is one of the means by which the JFC can shape the battle space, focusing on operational level objectives. It is designed to be used to attack the enemy simultaneously from all dimensions in a timed, cumulative manner that, when synchronized with other military activities, overwhelms his will to resist. Interdiction diverts, disrupts, delays, or destroys the adversary's surface military potential before it can be used effectively against friendly forces. Air interdiction is usually the largest contributor to the interdiction process, but other components, particularly special operation forces (SOFs), can also have a critical part to play. Air, land, maritime, and SOFs can conduct interdiction operations as part of their larger or overall mission.

At the operational level, maneuver is a means by which the JFC aims to set the terms of battle by time and location. This maneuver includes the decision to avoid battle or to adopt a defensive posture to impose offensive action at a time of his/her own choosing. Maneuver is the gaining of a position of advantage with respect to an adversary from which force can be threaded or applied; it is, in effect, the process by which combat power is employed to achieve a decision. In some circumstances, the psychological effect of maneuver can be so overwhelming as to render the adversary's subsequent actions futile; a bold and skilful maneuver at the outset of an operation may encourage the adversary to capitulate without ever fighting. Rarely, however, is this case. Maneuver can be accomplished by any military force element; but to be effective, it usually must be accompanied by the ability to apply force, usually in the form of potential fires to produce an operational and decisive effect.

Fires destroy, neutralize, suppress, and demoralize; it is the most violent manifestation of military force, and as such, it must be controlled to be effective. In joint operations, the synergy from mutually supporting fires from other components is known as joint fires, and it provides a powerful tool for the JFC to increase the effectiveness of his or her force.

2.3.2 Special Operations

SOF provides the JFC with a flexible, versatile, and unique capability, whether employed alone or complementing other forces or agencies, to attain military-strategic or operational objectives. Special operations, in contrast to conventional operations, are generally small, precise, adaptable, and innovative; they may be conducted in an overt, covert, or discreet manner.

Military activities conducted by specially designated, organized, trained, and equipped forces using operational techniques and modes of employment are not standard to conventional forces. These activities are conducted across the full range of military operations independently or in coordination with operations of conventional forces to achieve military, political, economic, and psychological objectives or a combination thereof. Political considerations may require covert or discreet techniques as well as the acceptance of a degree of physical and political risk not associated with conventional operations. Special operations may be conducted across the range of military operations as follows:

- SOF may be employed in support of the state's military-strategic objectives and operational objectives as directed by the JFC.
- SOF operations would be directed at the accomplishment of high-value, critical objectives that may entail high risk but also high pay-off value.
- Although SOF can be employed at the tactical level for a limited period of time (e.g., on a special, high-value task), these forces are limited in number, are not easily replaceable, and should not be used as a substitute for other, more appropriate forces.
- SOF can be most effective when employed during peacetime. They can contribute directly to enhance mutual cooperation, promote democracies, support peace operations, establish forward presence, provide early identification and assessment of a crisis, train friendly forces, or develop a military liaison. SOF may be employed in support of the information operations (INFO OPS) implementation at the strategic and operational level.
- In crisis, SOF can provide area assessments and an early C3 capability, complement and reinforce political activity, support the military response options, and assist in the transition from peace to crisis and conflict if necessary.
- On operations, SOF would conduct primary tasks of special reconnaissance and surveillance, military assistance, and direct action, or a suitable combination of tasks to support allied joint operations to repel aggression, restore peace, and assist in conflict termination; they would also assist postconflict activities during the transition back to peace.

In the context of joint operations, SOF conducts three principal tasks: special surveillance and reconnaissance, direct action, and military assistance. Special operations are marked by certain characteristics that distinguish them from conventional operations.

Special operations are characterized as follows:

- Usually of high physical and political risk and are directed at high-value, critical targets that offer the potential for high returns.
- Often politicomilitary in nature and require understanding, oversight, and full support at the national level. They demand detailed planning and coordination with other commands, services, government, and nongovernmental agencies.
- Usually conducted by small teams of highly trained, mature professionals.
- Generally dependent on responsive and specialized maritime, land, and air support.

- Usually covert or discreet in nature.
- Frequently undertaken when the use of conventional forces is, for military or political reasons, neither appropriate nor feasible.
- Reliant on surprise, security, and audacity; frequently employing deception to achieve success.
- Usually conducted at great distance from established support bases, requiring sophisticated communications and means of infiltration, exfiltration, and support to penetrate and recover from hostile, denied, or politically sensitive areas.
- May require patient, long-term commitment and detailed knowledge of the conditions, languages, and culture of the operational area.
- Dependent on discriminate and precise use of force, often requiring the rapid development, acquisition, and employment of special weapons and equipment.
- Dependent on detailed intelligence, intimate, and responsive C2, thoroughly planning decentralized execution and rigorous, detailed rehearsal.

2.3.3 Crises Response Operations

Crises response operations (CROs) range from support operations that are primarily associated with civil agencies through operations in support of peace, to tasks in support of disaster relief and of humanitarian, search and rescue, or noncombatant evacuation operations; evacuation of nationals remains a national responsibility. Operations that involve the use of military force or the threat of force include military action ranging from sanction and embargo enforcement to military combat operations. Furthermore, experience has shown that in addition to the availability of specialized units for a particular task, the military, because of its organization, capabilities, and ability to deploy rapidly, could be called on, in exceptional circumstances, to contribute to tasks that are the responsibility of mandated civil actors. Such tasks could run from public security and could engineer support for municipal services to border security/control.

The United Nations Security Council (UNSC) has the primary responsibility for the maintenance of international peace and security. The Organization for Security and Cooperation in Europe (OSCE) also plays an essential role in promoting peace and stability, enhancing cooperative security, and advancing democracy and human rights in Europe. NATO has offered to support, on a case-by-case basis in accordance with its own procedures, peacekeeping and other operations under the authority of the UNSC or the responsibility of the OSCE, including by making available alliance resources and expertise. All North Atlantic Treaty Organization (NATO) CRO have been under the political control and strategic direction of the North Atlantic Council (NAC), these operations were initiated by an NAC Initiating Directive and were executed in accordance with international law as well as international humanitarian law. CRO are politically sensitive and complex, and they demand close cooperation and coordination between the highest political and military bodies within the alliance as well as close consultation with partners when they are involved.

The operational environment in which CROs are taking place can range from permissive to hostile and will be influenced by the perception of the local population and local organizations, as well as the activities carried out on behalf of the International Community. The complex political and public nature of CRO should not be underestimated; however, these characteristics are exacerbated by the fact that the JOA may not be defined by clearly delineated boundaries and frontiers. Similarly, because of the difficulty of predicting how the situation may develop, a CRO may lack clear guidance on the end-state and criteria for success. The ability of the international community to react with resolve and in a timely manner is the key to success in conducting a CRO. This ability depends on a rapid international political and military decision-making process as well as the availability of appropriate military forces.

To have a mutual understanding of CRO and the relationship to reconstruction, it is essential to understand which players have a role in these operations (such as UN, NATO, European Union (EU), OSCE, nongovernmental organizations (NGOs), local structures, and many others). It is also necessary to examine closely the nature of the operation. Stabilization is the primary goal for development of the reconstruction efforts and ultimately normalization (Figure 2.1) [ETA05]. From the beginning of the operation, the coercion is on the highest level, where security and assistance to local government are on the lower level. By crossing to the stabilization phase, security in the

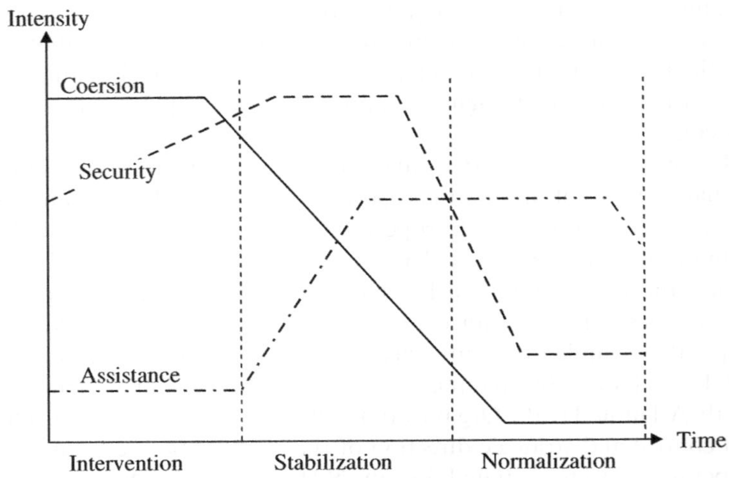

Figure 2.1: Military tasks in CRO.

affected society should be higher, coercion should be lower, and assistance should be on a balanced pace. The normalization phase is characterized by self-sustained local governance that could provide security for the local population. CROs are focused on achieving a safe and secure environment to proceed with reconstruction. CROs are not new: Many operations in which international military forces have participated, both recently and in the past, fall into this category. Recent events and overall geostrategic developments warrant a clear articulation of the principles for military forces engaged in CROs.

The stabilization phase is designed to build on initial military success and through all aspects of reconstruction with the goal of returning the area to normalcy. It is often conducted within a cultural environment that is extremely difficult for military forces involved to understand. It involves the intervention of many actors using varied strategies and with (political or institutional) interests that often diverge. Generally, it is a complex and fragile phase throughout; it can regress at any moment as a result of outbursts of violence. Furthermore, the progressive transfer of responsibilities from military to civil operators does not always happen at the same tempo within every functional area and therefore imposes a constant evolution on the way organizations function. It is characterized by a progressive involvement of all those actors who will contribute toward the achievement of normalization and ends as the normalization phase begins with the specific aim of the realization of a political objective, in turn permitting the complete withdrawal of the military forces. Conduct of the normalization phase is determined by how the stabilization phase has been executed.

Military activities, which are central to the success of the stabilization phase, are conducted within a joint civil-military environment in which civil actors—institutional and otherwise—are common. This environment is likely to be within a context that can be national but that is often multinational or international in nature, with large numbers of NGOs in the operating area. In most cases, it is conducted in a situation where most state functions are failing or are nonexistent. Military forces often have to deal with, directly or indirectly, the principal problems within the operating area; primary among these problems is security. This general approach implies a wide spectrum of activities that are conducted by military forces in liaison with all other actors involved in reconstruction efforts like defense and security, administration, economy, justice, health, education, and soon. The nature of challenges faced requires an ability to combine in space and time, military, humanitarian, and other activities.

2.3.4 Peace Operations

Operations in support of peace are recognized as an aspect of CROs; however, such operations are normally known as peace support operations (PSOs). They had been conducted impartially, normally in support of an internationally recognized organization such as the UN or the OSCE, involving military forces and diplomatic and humanitarian agencies. PSOs are designed to achieve a long-term political settlement or other specified conditions. They include peacekeeping and peace enforcement as well as conflict prevention, peacemaking, peace building, and humanitarian relief. The International Community responds to complex emergencies within a broad spectrum of activities according to the dimension of the crisis. Some development and relief agencies may have been operating in the area of the JOA for several years before the crises, and they will often play an important role as operations develop in an incremental and ad hoc fashion.

The success of a PSO is largely dependent on the support of the local population, their authorities where they exist, and the cooperation of the parties in conflict together with the International Community. The military task, in this context, will be to create the necessary security conditions and to provide support as necessary, so as to enable civil agencies to address the underlying causes of the conflict and thus generate a self-sustaining peace.

The size and military power of a "task force" involved in a peace operation depends on several factors, which are dictated by the security environment. The framework tasks for reconstruction should be organized around four distinct areas or *pillars*: security, justice/reconciliation, social/economic situation, and governance /participation [MAC96]:

- Security addresses all aspects of public safety, in particular the establishment of a safe and secure environment and the development of legitimate and stable security institutions. Security encompasses the provision of collective and individual security, and it is the precondition for achieving successful outcomes in other pillars. In the most pressing sense, it concerns securing the lives of civilians from immediate and large-scale violence and the restoration of territorial integrity.
- Justice and reconciliation addresses the need for an impartial and accountable legal system and for dealing with past abuses: in particular, the creation of effective law enforcement, an open judicial system, fair laws, humane corrections systems, as well as formal and informal mechanisms for resolving grievances that develop from the conflict. These tasks encompass the provision of mechanisms to redress grievances, appropriate penalties for previous acts of misbehavior, and a capacity to promulgate and enforce the rule of law. Incorporating the concept of restorative justice, they include extraordinary and traditional efforts to reconcile excombatants, victims, and perpetrators.
- Social and economic situation addresses fundamental social and economic needs: in particular, the provision of emergency relief, restoration of essential services to the population, laying the foundation for a viable economy, and initiation of an inclusive and sustainable development program. Often accompanying the establishment of security, well-being entails protecting the population from starvation, disease, and the

elements. As the situation stabilizes, attention shifts from humanitarian relief to long-term social and economic development.

- Governance and participation addresses the need for legitimate, effective political and administrative institutions and participatory processes: in particular, establishing a representative constitutional structure, strengthening public sector management and administration, and ensuring active and open participation of civil society in the formulation of government and its policies. Governance involves setting rules and procedures for political decision making and for delivering public services in an efficient and transparent manner. Participation encompasses the process for giving voice to the population through the development of a civil society that includes the generation and exchange of ideas through advocacy groups, civic associations, and the media [CSI02].

The execution of the above-mentioned tasks demands multidimensional synchronization of the capabilities of the civil and military components involved. Civil-military cooperation[3] in CEO can be introduced in many dimensions, such as follows: cooperation between the military component and local authorities or society; between local regular or unlawful forces and international civil organizations; and among military, police, and civil components. For CEO and reconstruction, this means cooperation between forces and the new political system, which has been created by the IC. These relations are complex, multidimensional, and in many cases undefined, which often lead to less efficient/effective military forces during reconstruction of the political structure in the affected area [MAR05].

CIMIC demands several security roles for involved components: 1) suppresses/eases tensions, 2) reconstruction of societal values, and 3) ensuring the origins for sustainable development of the affected society [EDM88]. Ramsbotham [RAM00] explained the sustainable developmental approach, which refers to causes of complex emergency, with the following definition: "In the context of approved measures by the UN, the enhancement of peace can be defined as a policy of external international help to the developing countries, assigned to support domestic social, cultural and economic development, and independence and self-sufficiency, to accelerate post-conflict recovery and reduction or cancellation of sources of violence in the future." Contemporary operations have been characterized by the close correlations between the decision-making process on the strategic level and the tactical execution of tasks, in terms of time and content. Thus, it is important to understand the prerequisites

[3] Civil military cooperation (CIMIC) is defined by the UN as follows: "The relationship of interaction, cooperation and coordination, mutual support, joint planning, and constant exchange of information at all levels between military forces, civilian organizations and agencies, and in theatre civil influences, which are necessary to achieve an effective response in the full range of operations."

on the tactical level for reaching the desired *strategic* end of the contemporary peace operation.

2.3.5 Network-Centric Warfare

Network-centric warfare (NCW) can be concisely defined as the concept of linking all aspects of war fighting into a shared situation awareness and understanding of command intent so as to achieve a unity and synchronicity of effects that multiplies the combat power of military forces. Network-centric operations are then military operations across the spectrum of conflict from peace, to crisis, and to war, to which the concepts and capabilities of NCW had been applied [SMI02]. The term "network-centric warfare" broadly describes the combination of emerging tactics, techniques, and procedures that a fully or even partially networked force can employ to create a decisive war-fighting advantage. NCW is warfare.

To understand what is different about NCW, as well as to understand the source of increased combat power associated with NCW, one has to focus simultaneously on the three domains of warfare and the interactions among them. These domains are the physical domain, information domain, and cognitive domain.

The physical domain is the place or location that the situation the military seeks to influence. It is the domain where strike, protect, and maneuver takes place across the environments of ground, sea, air, and space. It is the domain where physical platforms and the communications networks that connect them reside. Comparatively, the elements of this domain are the easiest to measure, and consequently, combat power has traditionally been measured primarily in this domain. In our analyses and models, the physical domain is characterized as reality or ground truth. Important metrics for measuring combat power in this domain include lethality and survivability.

The information domain is where information is created, manipulated, and shared. It is the domain that facilitates the communication of information among war fighters. It is the domain where the command and control of modern military forces is communicated and where the commander's intent is conveyed. The information that exists in the information domain may or may not truly reflect ground truth. For example, a sensor observes the real world and produces an output (data), which exists in the information domain. With the exception of direct sensory observation, all of our information about the world comes through and is affected by our interaction with the information domain. And it is through the information domain that we communicate with others. Consequently, it is increasingly the information domain that must be protected and defended to enable a force to generate combat power in the face of offensive actions taken by an adversary. And, in the all-important battle for information superiority, the information domain is ground zero.

The cognitive domain is in the minds of the participants. This is the place where perceptions, awareness, understanding, beliefs, and values reside and where, as a result of sense making, decisions are made. In this domain, many battles and wars are actually won and lost. This is the domain of intangibles: leadership, morale, unit cohesion, level of training and experience, situational awareness, and public opinion. Also in this domain exists an understanding of a commander's intent, doctrine, tactics, techniques, and procedures. Much has been written about this domain, and key attributes of this domain have remained relatively constant since Sun Tzu wrote "The Art of War." The attributes of this domain are extremely difficult to measure, and each subdomain (each mind) is unique.

Note that all the contents of the cognitive domain pass through a filter or lens that we have labeled "human perception." This filter consists of the individual's worldview; the body of personal knowledge the person brings to the situation; as well as their experience, training, values, and individual capabilities (intelligence, personal style, perceptual capabilities, etc.). Because these human perceptual lenses are unique to each person, we know that individual cognition (understandings, etc.) is also unique. Only one reality or physical domain exists; it is converted into selected data, information, and knowledge by the systems in the information domain. By training and shared experience, we try to make the cognitive activities of military decision makers similar, but they nevertheless remain unique to each individual, with differences being more significant among individuals from different services, generations, and countries than they are among individuals from the same unit or service.

2.3.6 Logistics

Effective logistic support is fundamental to the success of any operation, and therefore it must be an integral part of all operational planning. Although nations are ultimately responsible for the provision of resources to support their forces, commanders will ensure that the logistic force structure can support the operation and will coordinate support among contributing nations and with the host nation to ensure operational success. They also must be given, at the appropriate level, sufficient authority over the logistic resources necessary to enable the commander to employ and sustain the forces under command in the most effective manner.

Logistics encompasses the planning and carrying out of the movement and maintenance of forces, and it includes the disciplines of transportation as well as medical and health services. In its most comprehensive form, it also includes acquisition, supply and services, storage, distribution, equipment maintenance, evacuation, and disposal. In addition, the areas of infrastructure engineering and contracting are likely to be vital to the logistic effort. Cooperation, coordination, primacy of operations, flexibility, and synergy are the main principles that govern any joint and multinational logistic support planning. That is to say, the campaign logistic support concept must meet the mission and be flexible enough to facilitate a variety of national approaches, take advantage of national strengths, and indicate clearly that it is beneficial to national and international forces.

The logistic plan, structures, and procedures must be tailored to the respective forces and their related employment options. Logistic support options for a JFC can range from purely national support to multinational logistic support. The latter may encompass lead nation support, role specialization, and/or multinational integrated logistic support. Primary considerations in development of the logistic support structure include whether it is a combat or PSO; the type, size and scope of the mission; and host nation support (HNS) or contracting support availability in the JOA. Additional considerations include the following: 1) the extent of involvement by international organizations (IOS) and NGOs, 2) availability of bilateral acquisition and cross-servicing agreements, 3) the existence of additional requirements for rear area security, and 4) requirements for operating points of entry and lines of communication or for coordinating the use of real estate, contracting, and joint-force level engineering.

2.3.7 Information Operations

INFO OPS aims at achieving effects in the virtual and physical space in which information is received processed and conveyed. It consists of the information itself and information systems. These information systems comprise personnel, technical components, organizational structures, and information-based processes that collect, perceive, analyze, assess, create, manipulate, store, retrieve, provide, display, share, transmit, and disseminate information.

INFO OPS are defined as ‘actions taken to influence decision makers in support of political and military objectives by affecting other's information, information-based processes, command and control systems (C2), and communications and information systems (CIS), while exploiting and protecting one's own information and/or information systems. Two main categories of INFO OPS defensive INFO OPS and offensive INFO OPS are as follows:

- Defensive Information Operations. Defensive INFO OPS are defined as "actions taken to maintain access to and effective use of information, information based processes, C2 systems and CIS during peace, crisis or conflict and to protect force's information critical to achieving specific objectives". The conduct of force's defensive INFO OPS have to be build on security policies, doctrine, and procedures to ensure that information, information-based processes, and CIS are adequately protected against the effects of other's efforts to acquire, exploit, or otherwise manipulate information to the detriment of the force's military decision or political consultation processes and operational objectives.
- Offensive Information Operations. Offensive INFO OPS are defined as "actions taken to influence a potential adversary's available information, information based processes, C2 systems and CIS during peace, crisis or

conflict in pursuit of specific objectives or in reaction to a specific threat". The conduct of offensive INFO OPS will be based on political guidance from the military authorities' advice depending on a specific situation.

The application of INFO OPS can be considered at all levels of military operations—from military strategic to tactical. INFO OPS is concerned with the information objectives that a commander seeks to achieve by his actions. It is therefore a strategy that is both fundamental and central to a commander's planning of military activities. The JFC's campaign plan should be developed with the recognition that his designated JOA and beyond is an INFO OPS environment.

INFO OPS should be an integral part of the JFC's campaign planning from the outset and should indicate general operational objectives, be broad based, and encompass employment of all available capabilities—joint service, interagency, and multinational. INFO OPS planning should be conducted starting with the JFC's statement of intent. Planning should analyze the risk of compromise, reprisal, escalation of hostilities, and the uncoordinated or inadvertent counteraction of INFO OPS by the various joint service and/or interagency INFO OPS-capable providers that may be associated with a joint force.

An appropriate steering board should be created to coordinate, integrate and deconflict INFO OPS activities that contribute to fulfilling the JFC's intent (purpose, method, and end-state) within the JOA. This mechanism should be formed from representatives from each staff element, component, and supporting agencies responsible for integrating INFO OPS capabilities and related activities into the overall campaign plan.

2.3.8 Psychological Operations

Psychological operation (PSYOPS) is defined as follows: "planned psychological activities using methods of communications and other means directed to approve audiences in order to influence perceptions, attitudes and behavior, affecting the achievement of political and military objectives."

The psychological dimension of conflict is as important as the physical. Conflict is a struggle of wills that takes place in people's minds as well as on the battlefield. The attitudes and behavior of people (friend, foe, and the undecided or uncommitted) may be central to determining the outcome of conflict and the nature of the post-conflict environment. Therefore, it is necessary to understand the motivation of various leaders, military forces, and local populations to shape their perceptions, affect their will, and persuade them to accept the desired outcome by the IC. The employment of any element of power projection, particularly the military element, has always had a psychological dimension. PSYOPS, as a key element of most INFO OPS activity, is a vital part of the broad range of diplomatic, military, economic, and informational activities.

PSYOPS constitute a planned process of conveying messages to selected groups, which are known as target audiences, to promote particular themes that result in desired attitudes and behavior that affect the achievement of political and military objectives.

PSYOPS has the potential to affect an adversary's morale, negatively fear instill and breed distrust. It also has the potential for providing insights into adversary commanders' possible courses of action and can be used to direct other INFO OPS activities, such as deception, into areas where they are most likely to succeed.

PSYOPS policy identifies three categories for the conduct of psychological operations: strategic, crisis response, and combat. PSYOPS include strategic psychological activities (SPA), psychological consolidation activities (PCA), battlefield psychological activities (BPA), and peace support psychological activities (PSPA). Target audiences may be the military or civilian population of hostile or neutral nations, the adversary commander and staff, and friendly military or civilian populations. SPA and PCA are conducted at national government levels and do not normally involve military forces in their execution.

At the operational and tactical levels, BPA and PSPA normally involve military forces during both conflict and CROs. BPA and PSPA, which are conducted by strategic/regional commands and their subordinates, will not seek to target the international media, friendly nations or forces, or civilian audiences outside the JOA. By exception, PSYOPS units may provide support and technical assistance to non-PSYOPS activities, such as troop information or public information, but these units will not be given the responsibility to address such audiences. PSYOPS intent and activities vary with target audience.

The purpose of PSYOPS is to weaken the will of the adversary, bolster any support for allied objectives within the opposition's camp, and gain the support of the unsure or uncommitted by influencing the behavior and attitude of the adversarial audience. PSYOPS can be categorized by their apparent source: Black, Grey and White activities reflect messages that purport to emanate from a source other than a true one, do not specifically identify any source, or are disseminated and acknowledged, respectively, by the sponsor or by an accredited agency thereof.

PSYOPS are directed at the adversary's military audience with the aim of lowering morale; creating apathy, defeatism; and discord; and promoting dissension, subversion, uncertainty, defection; and surrender. Weak points in an adversary's political, economic, social; and military situations are identified and evaluated for importance, accessibility; and vulnerability. A coordinate, consistent attack is then launched at the adversary's political, economic, social, and military situations. These attacks are identified and evaluated for importance, accessibility, and vulnerability. The attack is delivered through the complementary use of various media and must be consistent with associated public affairs (PAS) activities and the strategic level information plan.

If planned and executed properly, the attack will create doubt or confusion in the minds of the adversary, and it will create doubt regarding the righteousness of their cause, the competence and integrity of their leaders, the dependability of their allies, the outcome of hostilities and, most important, the likelihood of their own survival.

2.3.9 Effect-Based Approach to Operations (EBAOs)

EBAO is an answer to the complexity of the global security environment and to the ever-changing political relations and interests in the IC. This term is not new and is related to the military desire to understand the operational challenges of the contemporary operations through by a more holistic and simplified language. According to Smith [SMI06], the strength of an EBAO is that it addresses these complexities squarely by concentrating on their most nonlinear aspects: humans, their institutions, and their actions. Indeed, the entire effects-based approach can be characterized by four things: a focus on the human dimension of competition and conflict; the consideration of a full spectrum of actions, whether in peace, crisis, or hostilities; a multifaceted, whole-of-nation concept of power; and the recognition of the complex interconnected nature of the actors and challenges involved. By writing that, he opened a "Pandora's box" of average society, where the interaction between different security dimensions shapes the capabilities of the local governance network. Basically, EBAO is allowing the explanation of objectives, operational means, actions, and tasks needed to accomplish desirable objectives.

The human dimension develops because all effects-based approaches are ultimately about shaping human perceptions and behavior, and because they depend heavily on human beings to make the complex estimates and decisions involved. Finally, any effects-based approach must proceed from the recognition that all actions and the reactions they provoke are inextricably linked in a system of ever changing and adapting human systems whose complexity shapes both the nature of the problem and the task of assessing, planning, and executing any operation.

With systemic theoretical modeling, it is possible to describe multilayered interactions on different societal levels. Each interaction can be described as an "action-reaction" cycle in which an individual person or organization reacts and adapts to a stimulus—anything from enemy fire to a diplomatic note. The stimulus enters the cognitive process through the eyes and ears of an observer who attempts to make sense of it, apply this understanding to judging options for a response, and choose a course of action or inaction that then becomes the end state of that cycle and the stimulus for a new cycle. But this time with the other side reacting in a continuing spiral of cycles, each of which builds on what has gone before and shapes those that will follow. Logically, each actor in the cycle would need to achieve some level of awareness of what was going on; to make enough sense of this picture to act or react; to decide on a course of action to deal with the challenges presented; to carry out those actions; and in doing all of this, to be subject to idiosyncratic social influences that shape their sense making and decisions [SMI06].

Good commanders have dealt with the complexities in these processes by inserting a "human in the loop"—whether themselves, subordinate commanders, staff, or watch personnel—to make complex decisions, to assess ambiguous information, and to fill in the blanks where information is wanting. In examining their real-world operations, we can identify where, why, and what human intervention was deemed necessary, as well as the generic problems, questions, and issues that the interventions addressed, and the requirements for information and knowledge that they engendered. Finally, the term "EBAO" has been used in a publications so far, to denote the diverse applications of effects-based thinking to operations, that is, approaches characterized by a focus on 1) the human dimension of collaboration, competition and, conflict; 2) the full peace-crisis-war-post-war spectrum; 3) integrated national or coalition power; and 4) the complex nature of problems and solutions. EBAO is a military, methodological contribution to the comprehensive approach of IC toward a certain area of complex emergency.

2.4 COMPREHENSIVE APPROACH TO OPERATIONS

At the center of the peace forces and others' preoccupations, the population constitutes a major player in stabilization. The support of the local population for the activities of military forces is therefore a prerequisite to success in stabilization. Hearts and minds must be won in the period immediately after the deployment of forces, which is a very small window of opportunity. The population, which is often in a state of shock and hoping for a radical change to the situation, is at best open armed and at worst expectant. In a situation where the social structures are often degraded and where the state apparatus is in decline, or has even disappeared, formal and informal networks develop and fill the administrative and security gaps. To increase efficiency of operations, we need to establish new methodical and applicable forms of cooperation, coordination, and analysis in the areas of complex emergency. The expected outcome includes recommendations for better results that are necessary for societal reconstruction of the affected society.

"Comprehensive approach" is term that recognizes the need to act together and refers to the collaborative context in which available stake–holders are used. This approach is best viewed as a braided rope, in which analysis, planning, decision making, and execution processes are entwined together from the beginning, and run through the entire spectrum of our effort both horizontally and vertically. This approach implies that each actor understands its capabilities and role in the overarching plan, where it is highly probable NATO will be the supporting rather than the supported command.

A comprehensive, systemic approach toward a complex emergency situation has enabled the IC to recognize the crisis area, gain an understanding of root causes, and implement the decision-making process for intervention and planning for execution of operation in a timely manner. It is absolutely critical to identify a proper methodology for the achievement of the desired end state of a return to normality in a more efficient way and ensure the future commitment of the International Community to solve challenges elsewhere. Normality in this context includes a sustainable security, reconciliation, and a structured society, which guarantees the basic needs of the local population.

The theoretical systemic model (Figure 2.2) [MAR05] shows three key systems that have an impact on CEOs: the international community, the area of complex emergency, and peace forces. International and national societal environments influence the structure and activities of peace forces. During decision-making processes concerning interventions, the security council (SC) considers the different options available as a security instrument of the UN, which often consists of military, police, and civil components. Military forces play an essential role in complex emergencies, not only for establishing a safe and secure environment but also for enabling other institutions to fulfill their duties as required. This demands a mutual understanding of each other's capabilities, strengths, and weaknesses. So far, experience has shown that cooperation among the different entities in a conflict area is not effective. A clear distinction regarding roles, tasks, and responsibilities does not exist, and

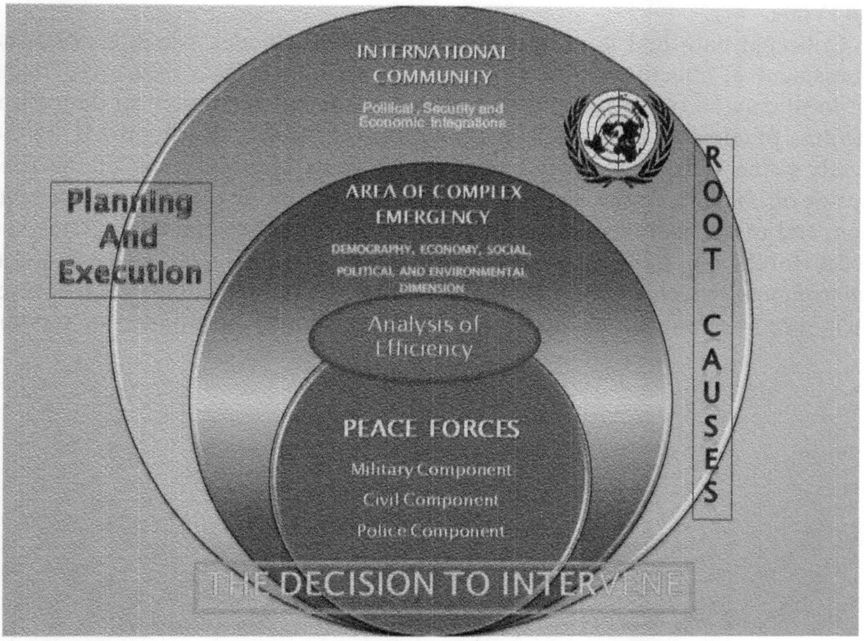

Figure 2.2: Theoretical systemic model of CEO.

an integrated body that could enhance cooperation well before a conflict emerges does not exist either.

Therefore, a more structured approach essential to achieving sustainable development for an affected society, based on a return to normality, which is the desired end state. A proper analytical methodology for better understanding of the area of CEO has become a necessity.

Multivariate analysis ensures a comprehensive and in-depth assessment of available indicators in the area of CEO. The methodology consists of systematic collection and analysis of empirical data about various security dimensions. The results of multivariate analysis explain the interactions and relations among structures, events, and processes.

To achieve desirable effects of CEO in the area of interest, it is necessary to coordinate local needs, international capabilities, and local capabilities for reconstruction. The need for societal reconstruction could be explained by the "triangle of societal reconstruction" (Figure 2.3), where societal reconstruction (SR) depends on international capabilities (IC) and local capabilities (LC), or with the descriptive formula SR = IC + LC. These three elements form the triangle of societal reconstruction in the area of CEO [DOY00].

With the good knowledge about societal security dimensions in the area of CEO (economy, demography, governance, landscape, and environmental issues) the International Community can monitor areas of complex emergency successfully through different phases: root causes, decision to intervene, planning and execution of CEO, and analysis of efficiency of CEO (Figure 2.2) [DOY00].

Collected data had been analyzed by a reactive model, which enables the assessment of effectiveness of peace forces, and it produces measured effects on local capability for societal reconstruction. Based on analytical data from the area of complex emergency, the systemic approach is steering the IC to the timely perception of complex emergency, toward understanding of the root causes of conflict, to the decision-making process about the type of intervention, and to the planning and implementation of the peace operation. Thus, the peace forces are tightly connected to the international and affected social environment, which is determining their composition and activities. Of course, their specific characteristics as typical hierarchical organizations or of systems

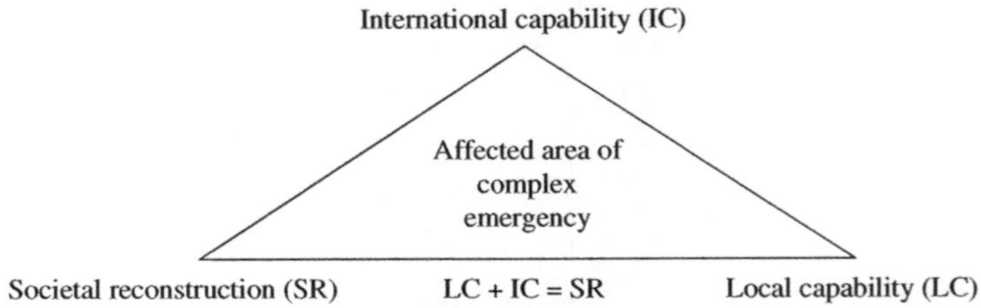

Figure 2.3: Triangle of societal reconstruction.

cannot be rejected, which include principle of subordination, command and control, training, logistics, and so on.

Despite the global dimensions of contemporary security, the national (state) dimension remains a key factor: The state ensures security to its own citizens with an active national security system. The effectiveness of this state is measured not only by the level of protection of its own fundamental societal values from external and internal threats, but also by the ability to provide economic, political, scientific, technological-technical, social, cultural, ecological, and other well-being issues for the population. Above all, the effectiveness of the state is measured by its ability to encourage sustainable development. The fundamental structural elements of a national security system are the operational capabilities of the society that can provide its own security. Contemporary political science has recognized the municipality/ province as the main local governance entity, which is needed to enable the overall societal security for its own population. Holistic societal analysis requires the proper selection of demographic, social, political, economic, and environmental variables at the municipality/province level to measure the potential of each security dimension and local capability for societal reconstruction as a whole.

2.4.1 Civil Military Cooperation

For abolishing the symptoms and resolving the causes of a complex emergency, it is necessary to develop a proper strategy and doctrine as an international answer that includes many different civil and military organizations. As a result, the peace operations are multipurpose, multidimensional, and multinational in that military forces are only one component in the operation with the shortest period of action in the area of complex emergency. Despite global dimensions of contemporary security, the national (state) dimension still stays as a key constant. On this level, the contemporary state ensures to HS own security citizens with an active national security system. The effectiveness of this system today shows not only an ability of the state, that it can protect its own fundamental societal values from external and internal threats, but also an ability to provide economical, political, scientific, technological-technical, social, cultural, ecological, and well-being to the population and, above all, sustainable development. The fundamental structural elements of the national security system are the operational capabilities of the society that can provide its security.

The characteristics of the intrastate conflicts in the 1990s in the world were absent from the security policy and disintegration of the security structure as a consequence of political differences. Thus, the main task of the IC today is the creation of suitable security mechanisms and instruments on the post-conflict area (Figure 6). The UNSC gives to the peace forces a mandate or operational demands by the resolution for a certain area of complex emergency. The components of peace forces are obliged to assist the local governance network in defining security principles and in reestablishing the national security system. The IC has no universally agreed, multilateral, or interdisciplinary concept of response to complex emergencies. Procedures change as the community crosses each new threshold of operational experience. The varying status and condition of the military forces, political leadership, and civil communities in the host state are the key factors that distinguish the different levels of operations. These variables dictate different responses. The size and military power of a response group involved in peace support operations sometimes has been increased, and its ability to enforce conditions has grown in a way that escalated the overall response. But as a rule, a response option is not altered except by a deliberate political decision that is supported by the necessary means to effect the operational change [MAC96].

Civil-military cooperation in peace operations can be introduced in many dimensions as follows: cooperation between the military component of peace forces and the local authority or society; cooperation between local regular or unlawful forces and international civil organizations; and cooperation among military, police, and civil components of peace forces. For peacekeeping operations, this means cooperation between peacekeeping forces and the new political system, which has been created by the International Community on the new bases. These relations are complex, multidimensional, and in many cases undefined, which often lead to less efficient peace forces during reconstruction of the political structure in an affected area.

The model of civil-military cooperation (Figure 2.4) shows a double security role for peacekeeping forces: 1) as a means for calming down the tensions, 2) as a means for reconstruction of societal values, and 3) as a means for the origins for sustainable development of a affected society [EDM88]. Ramsbotham [RAM00] explained sustainable developmental approach, which refers to the causes of a complex emergency, with the following definition: "In the context of approved measures of UN the enhancement of peace can be defined as a policy of external international help to the developing countries, assigned to support domestic social, cultural and economical development, an independence and self-sufficiency, to accelerate post-conflict recovery and reduction or cancellation of sources of violence in the future."

2.4.2 Economical and Social Aspects

As mentioned, most political conflicts in the last decades have been intrastate conflicts. The result was disintegration of political, social, and economic systems. A direct consequence was depopulation of crises areas on the countryside and demographic pressures on industrial centers by the affected local population. That result caused socioeconomic problems and deep societal regional crises. In many cases, no security systems exist, or the potential of local capability for societal reconstruction is very low. Thus, the primary task for the civil component of peace forces is to provide long-term conditions for development of political, social, and economical dimensions; the secondary

Figure 2.4: The model of civil-military coordination in the peacekeeping operation.

task is to ensure short-term help to the affected society. Mises [MIS01] has written that in war time, national economies are often excluded from the international economic events; they witness the destruction of material goods by military actions, increased consumption of military materials, and total loss of work force because of mobilization.

Poverty, low quality of living conditions, and malnutrition are common phenomena in the crises areas. Even the replacement for mobilized personnel with the females and youth does not return the level of precrises production. From an economical perspective, war and conflict are always bad for the economy, except for the post-war process of production, with the additional amount of goods to rebuild the nation, compensates losses of the war.

Galbraith [GAL73] has always warned that the role of the economy is to provide desirable products and services for the people. The population is in a prehistoric cycle for survival if nobody is providing basic needs for them.

The incorporation of a domestic economic system in global economic events is a sign that a normal economic situation in the country is in place. Main indicators for normalization are population needs, satisfied interests of big corporations, multinational companies, and chains of commercial enterprises, which include TV channels, banks, and state bureaucracy. All of these examples indicate long-term commitments for the IC to rebuild a post–conflict society.

Civil support in a peace operation could be classified into two phases:

Phase I—humanitarian assistance in emergency

Phase II—providing conditions for sustainable development

Smillie [SMI98] emphasized fact that civil organizations encountered three challenges by making effective links between relief and development: timing, funding, and understanding. The first challenge concerns appropriate timing—when to engage, when to modify the intervention, and when to withdraw. As essential ingredient of prevention, conflict reduction, and conflict resolution, appropriate timing is important in terms of knowing if, when, and how to move from basic humanitarian relief to more developmental objectives. Examples of the problem include action too late in Rwanda and perhaps Kenya; departure too soon in Haiti; and transition too fast in Cambodia, Sierra Leone, and Bosnia. Key determinants of timing include political will and financial resources. They can be implicated in hurrying or delaying humanitarian response as well as in rushing the move from relief to development programming. They can be the cause of precipitous agency withdrawal and of the recent obsession with "exit strategies," which may or may not be appropriate to the pace of social and economic change on the ground.

Funding is the second challenge. Emergency funding remains sporadic, arriving in short-term bursts and often after lengthy delays. It can be patchy, and much of it is overtly political. Development assistance too can be patchy, cumbersome, and rigid; it often arrives late and without reference to the emergency that it follows.

Knowledge and information are related but different; together they characterize the third and most important challenge: understanding. Understanding represents the most difficult challenge, a sine qua non for proper timing and a prerequisite for the wise use of whatever funding is available. Although both knowledge and information may be in short supply, much greater emphasis has been placed by practitioners on information, especially at either end of the relief-development spectrum, than on knowledge. Inappropriate blueprint-type reconstruction and rehabilitation programs continue to abound, in part because of serious impediments to institutional learning. These impediments include a fear of, and a consequent aversion to, evaluation and an environment in which relief workers suffer from danger, stress, overwork, and burnout. These realities leave institutional memories shallow and provide experienced workers with inadequate time to educate others.

The effectiveness of civil organizations could be assessed by the direct effects in the areas of responsibility. Indicators of their success should be viewed in a decreased number of refugees and internal displaced persons, in improvements of medical conditions for the affected population, in improvements of housing, infrastructure, and in assistance with the inclusion of demobilized persons to civil society, and so on. The more difficult part is development of industrial capabilities, new jobs, and growth of incomes. It is necessary to decrease inflation and unemployment, increase gross domestic product (GDP), rebuild degraded economic system of the affected country, and incorporate it in international commerce.

If humanitarian assistance is a tactical level of effect, the restoration of the economy is an operational and strategic effect on the reconstructed society, which is often linked to local, regional, and state political authorities.

2.4.3 Comprehensive Approach and Its Application

The application of a comprehensive approach to contemporary operations is in the planning phase, execution phase, and analysis phase. As explained, the holistic approach is important to the interactions among local capability is for peace of the affected society, international capabilities for assistance, and required activities for initiation of sustainable development.

With the comprehensive systemic approach, we have to understand the initial potential of societal security dimensions in the area of complex emergency. Thus, for proper risk and capability assessment of local authorities, it is necessary to determine functions and societal areas for analysis. They could be connected with the object of risks, areas of risks, and how the local authorities are dealing with possible threats and developing their own skills for decreasing vulnerabilities. The vulnerability of the local community is in correlation with its capabilities to manage complex security conditions and, therefore, the level of survivability for the population. Local authorities with an acceptable level of vulnerability could be defined as capable and flexible. The most appropriate method is a multivariate analysis of local capability with a cluster analysis and factor analysis, which allows the researcher to define typology and clusters of municipalities/provinces by groups of variables (similarities).

Example 2.1

As a case study, Afghanistan was investigated by its local capabilities for societal reconstruction in 34 provinces at the beginning of 2007. Variables for secondary analysis were collected from open sources on the Internet, as well as from the Statistical Office of Afghanistan, the Afghanistan Information Management Service, International Security Assistance Force Headquarter (ISAF HQ), United Nations High Commissioner for Refugees (UNHCR), UN Office on Drugs and Crime, and some other humanitarian organizations. For populating the multivariate analytical model, the province was used as basic, two-dimensional statistical units that were defined by the name and size of the area. All 34 provinces were compared by 42 variables structured as follows: 9 demographic, 10 social, 11 economic, 6 political, and 6 environmental variables. With the help of statistical computer software SPSS 11.0 (SPSS Corporation, Chicago, IL) it was possible to compare similarities among provinces by descriptive and numerical variables.

Cluster analysis produced a clear picture of three distinctive groups of provinces (Graph 2.1): one (Kabul) very capable, 4 (Balkh, Herat, Kandahar, and Nangarhar) capable, and 29 less capable for self-sustained societal reconstruction (Graph 2.1).

Kabul had the highest level of measured variables above statistical average, which is twice as high as the second capable group of four provinces. Overall, 85% of the provinces in Afghanistan showed levels of societal dimensions below statistical average. They would require additional international capability for sustainable development.

With the cross tabulation of two variables, Province and local capability for peace (Table 2.1), it was obvious which province belonged to which distinctive group. These results can be visualized by the global information system (GIS) layer (Figure 2.5). Cross tabulation allows the researcher to compare local capability with the single indicator of societal dimensions to produce short-term planning for peace forces activities. For long-term planning, the factor analysis (Figure 2.6) showed two main factors or latent variables, which include Societal development and black economy and insurgency.

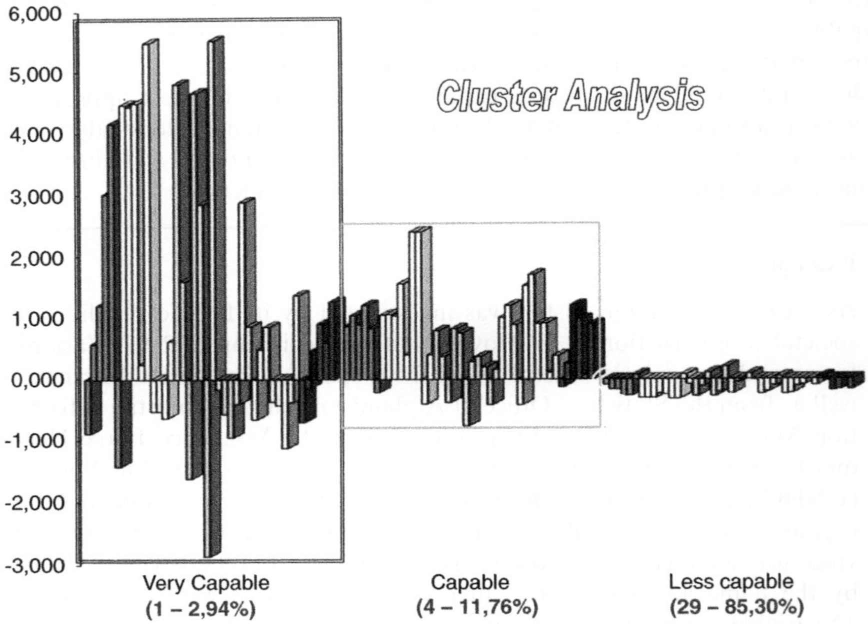

Graph 2.1: Local capability for peace in Afghanistan.

Table 2.1: Cross tabulation of province and local capability for peace.

		Local capability for Peace			
		Very capable	Less capable	Capable	Total
Province	Badakkhistan		1		1
	Badghis		1		1
	Baghlan		1		1
	Balkh			1	1
	Bamyan		1		1
	Day Kundi		1		1
	Farah		1		1
	Faryab		1		1
	Ghazni		1		1
	Ghor		1		1
	Helmand		1		1
	Herat			1	1
	Jawzjan		1		1
	Kabul	1			1
	Kandahar			1	1
	Kapisa		1		1
	Khost		1		1
	Kunar		1		1
	Kunduz		1		1
	Laghman		1		1
	Logar		1		1
	Nangarhar			1	1
	Nimroz		1		1
	Nooristan		1		1
	Paktika		1		1
	Paktya		1		1
	Panjshir		1		1
	Parwan		1		1
	Samangan		1		1
	Sar-i-Pul		1		1
	Takhar		1		1
	Uruzgan		1		1
	Wardak		1		1
	Zabul		1		1
Total		1	29	4	34

Cross tabulation among local capability and the two most influenced factors is explained by 43.9% of variance, which shows ways for the IC to address the complex situation correctly in Afghanistan. Kabul is clearly the most developed province in the country. The intention of the IC should be to bring most provinces to the top left quadrant with the **high societal development** and **weak black economy and insurgency**. Recent research of public opinion made by the

Figure 2.5: Regional reactive analytical model of Afghanistan.

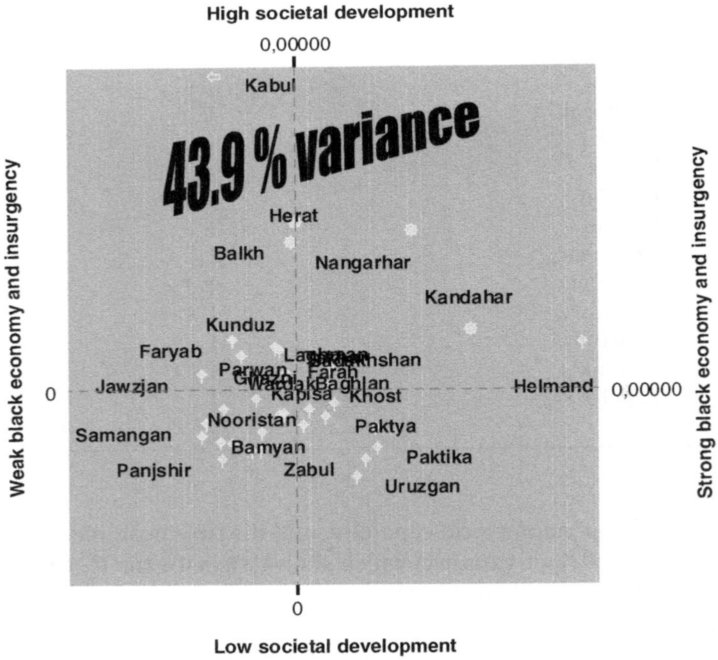

Figure 2.6: Factor analysis.

UNODC [UNO07] in 508 villages in all 34 provinces discovers that in 95.3% cases, the main reasons for a black economy are linked to the social and economic dimension. Local authorities are not capable of dealing with the severe poverty in the country.

With this kind of comprehensive approach to the area of complex emergency, it is possible to react in a timely manner and with appropriate security instruments and measures to increase the capabilities of the IC and to assist the affected society.

2.5 REVIEW QUESTIONS

2.1 Describe the types of conflicts.
2.2 What are the basic differences among theories of war?
2.3 Compare generations of warfare explained by Toffler.
2.4 What are the key concepts of operation?
2.5 What is the battle space?
2.6 Define crises response operation.
2.7 Explain the framework tasks for reconstruction of the affected society.
2.8 Which are the three domains of network-centric warfare?
2.9 Why is logistics so important?
2.10 Explain the two main categories of an information operation.
2.11 What is the strength of the EBAO?
2.12 What is achievable by the comprehensive approach to an operation?
2.13 Classify the civil support in a peace operation.

3

STATISTICS AND PROBABILITY

After discussions on possible approaches of the international community (IC) to the areas of complex emergency, nature of military and civil operations, and military functional areas, which can be exercised during computer-assisted exercise (CAX), it is time to explain how to make a combat model for requested training conditions. Statistics, probability, and randomness are the essence of combat modeling. A good understanding of the fundamentals related to these topics is essential not only for model designers, implementers, developers, and operational research analysts but also for exercise controllers and decision makers that use these tools especially for defense planning, supporting operations, and acquisition purposes.

A combat model is the representation of a real-world phenomenon by using many simplifying assumptions, which are generally in the form of randomness. A person who does not have an insight about statistics, probability, and randomness cannot assess well enough the implications of the results by a combat model. Therefore, this Chapter is dedicated to these topics. For readers from an operations background and not technical, this chapter provides fundamentals. For technical readers, it is a summary of the essentials. We do not intend to provide a complete course material for statistics in this chapter. Instead, we summarize the topic and provide examples more related to the military domain and reference material needed for the rest of the book.

Computer Assisted Exercises and Training: A Reference Guide, by Erdal Çayirci and Dusan Marincic

3.1 DESCRIPTIVE STATISTICS: POPULATION, SAMPLE, CENTRAL TENDENCY, AND DISPERSION

Statistics provide an insight into population, which includes all the elements of a phenomenon whose characteristics are studied. Because a population is often too large and not available as a whole, typically a subset of the population is selected and examined for statistics. This subset is called a sample. A **statistic** is a value for a property of a sample. A Sample statistic can be used for estimating a population parameter, which is a value for a property of a population. The larger the sample is, the more accurate a sample statistic becomes in estimating a population parameter. Not only the size but also the selection of the sample is important. The sample should not be selected from a certain part of the population because it causes a bias to that part in the result, and the sample statistic may not be accurate enough in representing the related population parameter.

Example 3.1

Population: The result of launching a Type X anti-tank rocket.

Sample: Because it is not possible to record the results of all rocket launches, n number of recorded results are selected for statistics. These results are our sample.

Population parameter: The ratio P_h between the hits and the launches for anti-tank rocket of Type X.

Sample statistic: The ratio S_h between the hits and the launches in our sample of n records.

Examples for biased sample:

- The results for the rocket launched at night.
- The results of the rockets fired against static targets.
- The results obtained in a firing range.

Several methods are used to collect a sample out of a population. We can categorize these techniques into three broad classes as follows:

- Random sampling
- Sample of convenience
- Judgment sample

Random sampling can also be classified as simple random, systematic, and stratified sampling. When each element of a population has an equal chance of being selected, it is called simple random sampling. In systematic sampling, every n_{th} element of a population is selected for the sample. For stratified sampling, the population is first divided into logical groups called strata, and then a sample is taken from each stratum. In a sample of convenience and a judgment sample, the elements of a sample are not selected randomly. A sample of convenience is an already available sample, and a judgment sample is selected by an expert about the population.

Statistics can be divided into two broad areas as descriptive and inferential statistics. Descriptive statistics includes organizing, presenting, and describing sample data, which are collected from a population. Inferential statistics use the results from a sample to estimate the characteristics of the population. We will elaborate inferential statistics later. For descriptive statistics, various ways are employed to organize, present, and describe data, which can be qualitative or quantitative. Examples are as follows:

- Data array
- Stem and leaf
- Frequency distribution
- Histogram
- Percentile

Example 3.2

a. Data array for the ages of the personnel in Platoon X:

27, 19, 20, 20, 21, 21, 20, 20, 20, 21, 25, 25, 21, 20, 20, 20, 21, 25, 25

b. Stemplot for the data array in Example 3.2.a

15	4
20	0 0 0 0 0 0 0 0 1 1 1 1 1
25	0 0 0 0 2

Key 15|4 means 19 years old
20|0 means 20 years old

c. The marks of 20 students in a test
16, 23, 28, 32, 34, 43, 47, 52, 52, 55, 56, 57, 61, 73, 76, 77, 77, 85, 86, 94

d. Stemplot for the data array in Example 3.2.c.

1	6
2	3 8
3	2 4
4	3 7
5	2 2 5 6 7
6	1
7	3 6 7 7
8	5 6
9	4

Key 1/6 means 16 marks

Data arrays as well as stem-and-leaf diagrams (stemplots) are two basic ways of organizing data. Stemplots are also useful to observe how the data are distributed among selected intervals. For stemplots, the data are organized first into groups, and the first value in the group gives the stem; then the number added to get the actual data point becomes the leaf as shown in Example 3.2.b. Finally a key is used to explain what the stem and leaf indicates. Another approach for stemplots is shown in Example 3.2.d, where the first digit of each data item gives the stem, and the second digit is used as a leaf value.

A Frequency distribution of data can be examined by listing all the data values in the sample together with the number of times that the same data values are observed, (i.e., frequency of the data value). The results can be presented through a frequency table.

Example 3.3

Frequency table for Example 3.2.a:

Age	Frequency
19	1
20	8
21	5
25	4
27	1

Grouped frequency distribution is an approach where the data are grouped into classes with equal intervals called class width. The midpoint of each class represents the class, and each class is called a class mark.

Example 3.4

Grouped frequency table for Example 3.2.a:

Class width = 5

Class mark	Class boundaries	Class limits	Frequency
18	15,5–20,5	16–20	9
23	20,5–25,5	21–25	9
28	25,5–30,5	26–30	1

Relative frequency can be more descriptive than frequency. A relative frequency is the ratio between the frequency and the total number of observations in the sample, (i.e., sample size). Note that the sum of relative frequency values should be equal to one if everything is calculated correctly.

Example 3.5

Frequency table for Example 3.2.a:

Sample size $n = 19$

Age	Relative frequency
19	1/19 = 0.0526
20	8/19 = 0.4211
21	5/19 = 0.2632
25	4/19 = 0.2105
27	1/19 = 0.0526
	Sum = 1.00

A frequency table can also be depicted by using various forms of graphs, such as a line graph, bar graph, and pie chart. A bar graph for a frequency table is called a histogram, and a line graph is called a frequency polygon. Histograms and polygons give an idea about the shape of the distribution of sample data. If the graph is not symmetric in shape, it is called skewed distribution. For example, the distribution of data in Example 3.6 is skewed. Please note that frequency polygon and histogram are depicted in Example 3.6 (Figure 3.1 and 3.2) A relative frequency polygon and histogram can also be shown by the same type of graphs. See Figure 3.3 for an example of a frequency pie chart.

Example 3.6

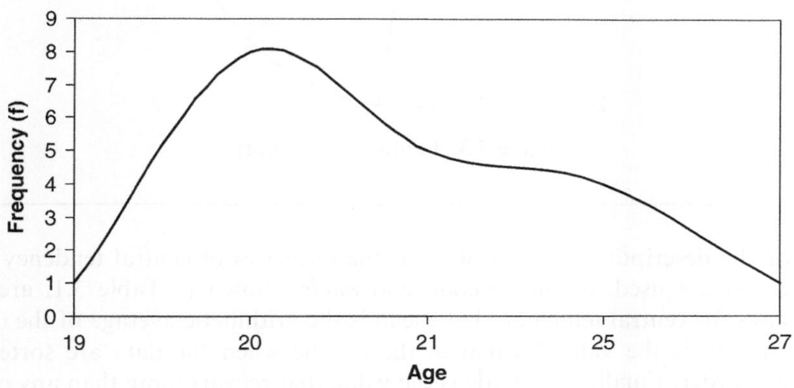

Figure 3.1: Frequency polygon.

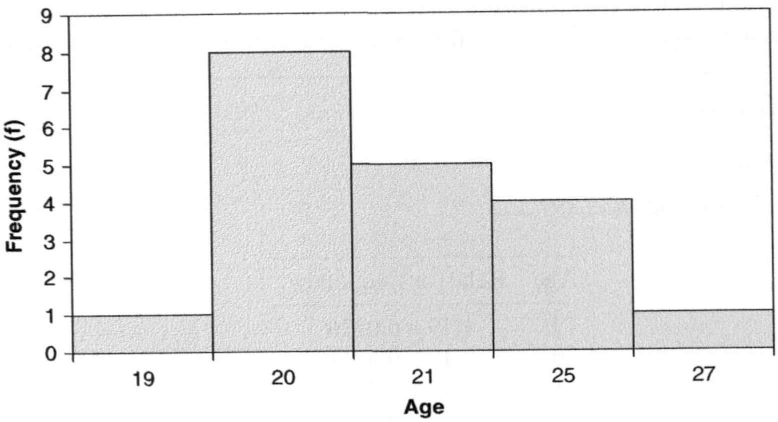

Figure 3.2: Frequency histogram.

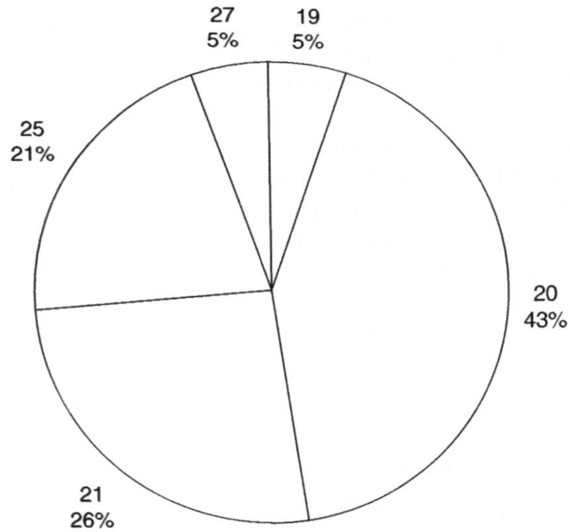

Figure 3.3: Frequency pie chart.

For the descriptive analysis of data, the measures of central tendency and dispersion are used. ***Mean***, ***median***, and ***mode***, shown in Table 3.1, are the measures for central tendency. The mean is the arithmetic average of the data. The median is the value located in the middle when the data are sorted in numeric order. Finally, the mode is the value that repeats more than any other value in a data set.

Table 3.1: The measures to describe central tendency.

Term	Population value	Sample value	Description
Mean	$\mu = \frac{\sum_{i=1}^{N} x_i}{N}$	$\overline{x} = \frac{\sum_{i=1}^{n} x_i}{n}$	Average value.
Median	T	M	The value located in the middle of a sorted data array.
Mode	—	—	The most frequent value.

Table 3.2: The measures to describe dispersion.

Term	Population value	Sample value	Description
Range	max – min	max – min	Difference between maximum and minimum values
Variance	$\sigma^2 = \frac{\sum_{i=1}^{N} x_i - \mu}{N}$	$s^2 = \frac{\sum_{i=1}^{n} x_i - \overline{x}}{n-1}$	The average of the squared distance of the values from the mean
Standard deviation	Σ	S	Square root of variance

Example 3.7

Mean median and mode for the data in Example 3.2.a:

Mean: $\overline{x} = \frac{411}{19} = 21.6316$
Median: $m = 21$
Mode: mode = 20

Range, variance, and standard deviation, which are shown in Table 3.2 are the measures for dispersion. Range is the difference between the largest and smallest values in a data set. Variance and standard deviation indicate the average difference between the values in the data set and the average of the data set. Variance is the average of the squares of the differences between each value and the average. Standard deviation is the square root of variance.

Example 3.8

Range variance and standard deviation for the data in Example 3.2.a:

Range: $r = 27–19 = 8$
Variance: $s^2 = \frac{104.42}{19} = 5.5$
Standard deviation: s = 2.34

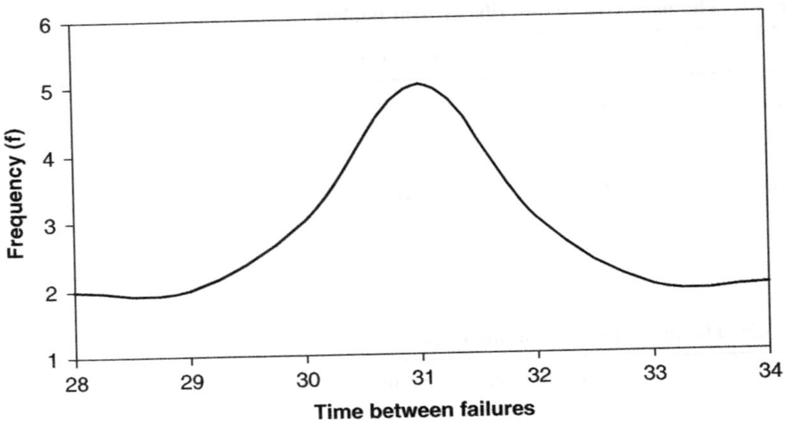

Figure 3.4: Time between failures for Type *Y* radio.

If the data are not skewed, then its frequency polygon has a bell shape, and the data have a normal distribution. In a normal distribution, data may still have dispersion, (i.e., range, variance, and standard deviation are greater than 0). We will elaborate distributions, normal distribution, and inferential statistics after probability and randomness in this chapter.

Example 3.9

Time in days between the failures of Type Y radio in combat: Figure 3.4
31, 30, 33, 30, 31, 28, 29, 28, 32, 31, 34, 32, 29, 31, 30, 34, 31, 33, 32

Mean: $\overline{x} = \frac{589}{19} = 31$
Median: $m = 31$
Mode: mode $= 31$
Range: $r = 34 - 28 = 6$
Variance: $s^2 = \frac{58}{19} = 3.0526$
Standard deviation: $s = 1.7472$

Please note in this example, that mean, median, and mode are equal to each other.

3.2 PROBABILITY

Probability is the likelihood of the occurrence of a simple event, which is a basic outcome of an experiment. An experiment is the process of obtaining an observation or taking a measurement. In a classic approach, probability indicates the relative frequency of a simple event (i.e., the ratio between the number of observations of a simple event and the total number of outcomes in the related sample, which is the collection of all simple events of an experiment). For example, let A represent a simple event. The probability $P(A)$ that A occurs is as follows:

$$P(A) = \frac{\text{number of ways that } A \text{ happens}}{\text{number of outcomes in sample } S} = \frac{n(A)}{n(S)} \tag{3.1}$$

Two important rules for assigning probabilities to simple events are as follows:

- All simple event probabilities must lie between 0 and 1.

$$0 \leq P(Ei) \leq 1 \quad \text{for all } i$$

- The sum of probabilities of all simple events within a sample space must be equal to 1.

$$\sum_{i=1}^{k} P(E_i) = 1$$

Finally, an **event** is a specific collection of simple events, and the probability of an Event X is equal to the sum of the probabilities of the simple events in Event X.

Example 3.10

What is the probability that there are two machine guns in a team of three troops, if each troop randomly picks either a machine gun or a rifle, that is, the probability that a troop picks a machine gun is equal to the probability that the troop picks a rifle?

A: Two troops select a machine gun and the other selects a rifle
M: A troop selects a machine gun
R: A troop selects a rifle
S: All possible outcomes
S: {MMM, MMR, MRM, MRR, RMM, RMR, RRM, RRR}
n(A) = 3
n(S) = 8

$$P(A) = \frac{n(A)}{n(S)} = \frac{3}{8} = 0.375$$

3.2.1 Counting Techniques

One approach to finding out the probability that a simple event occurs is counting. One technique for counting is a tree diagram in which all the possible outcomes are represented for each occurrence. This approach helps to find the number of outcomes in a sample space and the number of ways an event occurs. In Example 3.11, the sample space for the outcomes of tossing a coin three times is depicted by using a tree diagram. Because eight outcomes are possible for tossing a coin three times as shown in, Figure 3.5 in Example 3.11, the probability of each combination, (e.g., a tail (T) followed by a head (H) followed by another tail (T)) is 1/8. A counting tree can easily become too cumbersome as the number of possible outcomes for each trial and the number of trials increase. For example, to draw a counting tree for rolling a die for seven times is not easy.

Example 3.11

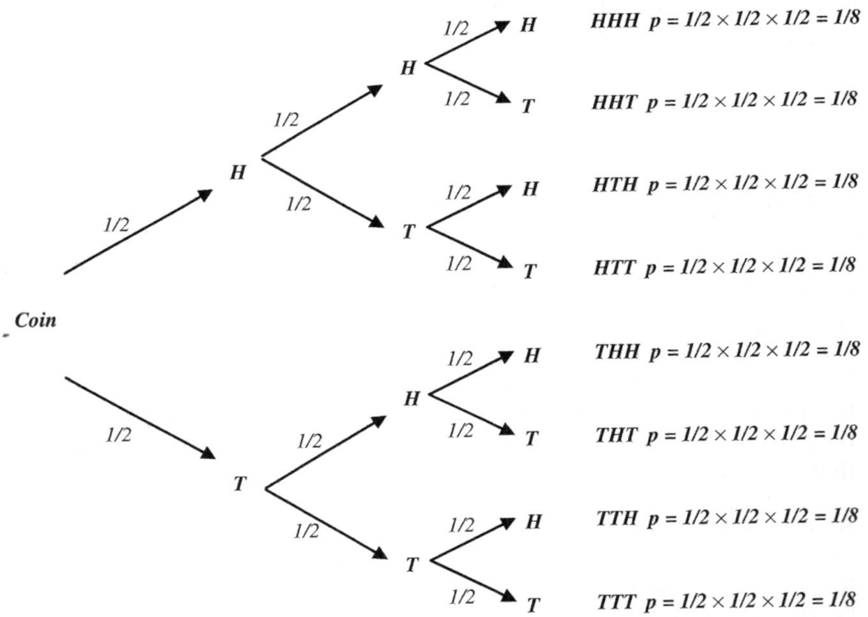

Figure 3.5: Sample space for the outcomes of tossing a coin three times.

To simplify the counting process, a ***counting rule*** can be used. Let us assume that we have three parts in our experiment, the number of possible outcomes (i.e., events) for the first part is k, the number of possible outcomes for the second part is m, and the number of possible outcomes for the third part is t. Then, $k \times m \times t$ outcomes are possible for this experiment. In Example 3.11, there are three trials of tossing a coin, which has two possible outcomes at each trial. Therefore, the total number of outcomes for tossing a coin three times is $2 \times 2 \times 2$, which is equal to 8, and the probability of one specific combination of outcomes is 1/8. A counting rule provides a way that finds out the number of all possible outcomes and the probability that a certain combination of outcomes may occur. For a counting rules it is important to calculate the number of all possible outcomes. Example 3.12 gives two counting problems and their solutions.

Example 3.12

a. In a multiple-choice examination of 10 questions that have four choices (i.e., one correct and three wrong answers)
   - How many ways can be used to answer the questions?
   - How many ways can be used to answer all the questions wrong?
   - How many ways can be used to answer all the questions correct?

   There are $4 \times 4 \times 4 \times 4 \times 4 \times 4 \times 4 \times 4 \times 4 \times 4 = 4^{10}$ ways to answer the questions.

   There are 3^{10} ways to answer all the questions wrong.

   There is only one (1^{10}) way to answer all the questions correct.

b. In a battalion, there are one lieutenant colonel (LTC), four majors (MAJ), and eight captains (CPT). How many ways can be used to pick one LTC, one MAJ, and one CPT?

   There are $1 \times 4 \times 8 = 32$ ways to pick one LTC, one MAJ, and one CPT.

Permutation and combination rules can also help for counting the possible outcomes. When not only the content but also the order of outcomes matter, a permutation can be applied for counting. When only the content, (i.e., not the order) matters, a combination can be applied for counting. Let us assume that we have two things, and there are k ways of doing them. Then, there are $k \times k$ outcomes. This example is a counting rule. However, when we use one way of doing things only once (i.e., when one way is used for one thing, then it cannot be used for the other), then there are $k \times (k-1)$ outcomes. This rule is a permutation, where applying one way to one thing or to the other are counted as different outcomes. When we do not count them as different outcomes, then it becomes just picking two ways out of k ways, which means that $(k \times (k-1))/2$ outcomes are possible. Permutation and combination rules are explaine in more detail in Example 3.13.

Example 3.13

a. There are two posts and five people. How many different ways can be used for the assignment to these posts?

   We can assign one of the five people to the first post. Therefore, there are five different ways for the assignment to the first post. However, there will be only four different ways for the second post because one will be already assigned. Hence, there will be $5 \times 4 = 20$ different ways for this assignment. This solution can be formulized by permutation as follows:

$$P_n^N = \frac{N!}{(N-n)!} \qquad N = 5, n = 2$$

$$P_2^5 = \frac{5!}{(5-2)!} = \frac{5 \times 4 \times 3 \times 2 \times 1}{3 \times 2 \times 1} = 20$$

Note that in this answer we consider posting Person X to Post 1 as a different outcome from posting the same Person X to Post 2.

b. There are two posts and five people. How many different ways are there for picking two from these five people for these two posts? Note that we do not count posting a person to Post 1 or Post 2 as different options. Therefore the problem is reduced to a problem of picking two out of five people. We can solve this by a combination:

$$C_n^N = \frac{N!}{(N-n)! \times n!} \qquad N = 5, n = 2$$

$$C_2^5 = \frac{5!}{(5-2)! \times 2!} = \frac{5 \times 4 \times 3 \times 2 \times 1}{(3 \times 2 \times 1) \times (2 \times 1)} = 10$$

c. How many different ways can be used to permute two letters from ABC?

AB, AC, BA, BC, CA, CB

$$P_2^3 = \frac{3!}{(3-2)!} = \frac{3 \times 2 \times 1}{1} = 6$$

Note that we do not count AA, BB, and CC as possible outcomes.

d. How many different ways are there to combine two letters from ABC?

A and B, A and C, B and C

$$C_2^3 = \frac{3!}{(3-2)! \times 2!} = \frac{6}{2} = 3$$

Let us make the problem in Example 3.13.b and d a little more complex. What if we need to assign more than one person to each of the posts or what if we have the right to use each letter multiple times? These problems can be solved by using another approach called partitioning as shown in Example 3.14.

Example 3.14

a. There are three posts and 16 people ($N = 16$). For the first post, three people ($k = 3$), for the second post four people ($m = 4$), and for the third post nine people ($j = 9$) need to be assigned. How many different ways are there for these assignments?

$$S = \frac{N!}{k! \times m! \times j!} = \frac{16!}{3! \times 4! \times 9!} = 400,400$$

b. How many different nine-letter ($N = 9$) combinations can be made by using two Bs ($k = 2$), three Cs ($m = 3$), and four As ($j = 4$)?

$$S = \frac{N!}{k! \times m! \times j!} = \frac{9!}{2! \times 3! \times 4!} = 1260$$

When the number of possible outcomes $n(S)$ in a sample and the number of outcomes $n(A)$ that represent an event are counted, the ratio between $n(A)$ and $n(S)$ gives the probability $P(A)$ for the event. We already explained this in the beginning of this chapter. Let us list the steps for calculating the probability of an event as follows:

- Define the experiment
- List all the simple events
- Assign probability to each simple event
- Determine the collection of simple events contained in the event of interest
- Sum the probabilities of simple events in the collection to find out the probability for the event

Observe these steps in the following example:

Example 3.15

Experiment: What is the probability that all three questions that have four multiple choices are answered correctly?

All possible simple events: There are 4^3 ways of answering three multiple-choice questions that have four choices.

The event that every question answered correctly: *There are 1^3 ways of answering all questions correctly. The probability* $P(A)$ that all three questions are answered correctly when one answer is picked randomly for each question:

$$P(A) = \frac{1^3}{4^3} = \frac{1}{64} = 0.015625$$

The next topic in probability is a concept called complement. The complement of an event is the set of all the other events in the related sample, and, therefore, the probability that the complement of an event happens can be formulized as

$$P(\overline{A}) = 1 - P(A)$$

For example, the probability for the complement of the event that all the answers are answered correctly in an exam of three questions with four choices is

$$P(\overline{A}) = 1 - 0.015625 = 0.984375$$

The formula for a complement can be very useful for many cases as in this example. For example, finding the probability that at least one question is answered wrongly requires calculating the probability for one, two, and three wrong answers and summing the results. This method can be too cumbersome especially when the number of questions increases. However, the event that there is at least one wrong answer can be stated also as the complement of the event that all the answers are correct. Of course the former approach is much easier to compute as we already did above.

3.2.2 Independence, Multiplication Rule, and Conditional Probability

Another concept in probability is the independence of an event. If an event does not have any influence in the occurrence of another event, then those two events are independent from each other. Otherwise, the events are dependent. This approach can be stated more formally as "the probability $P(A)$ of an event A occurs is not dependent on another event B occurs, therefore the probability of A occurs knowing that B has occurred $P(A|B)$ is not different from $P(A)$." Hence, two events A and B are independent from each other only if the following are true:

$$P(A|B) = P(A) \text{ and } P(B|A) = P(B)$$

When two events A and B are independent from each other, the probability that both A and B occurs $P(A \cap B)$ is equal to the multiplication of $P(A)$ and $P(B)$.

$$P(A \cap B) = P(A) \times P(B)$$

This rule is called the multiplication rule. When two events A and B are dependent, the multiplication rule becomes

$$P(A \cap B) = P(A) \times P(B|A)$$

$$P(A \cap B) = P(B) \times P(A|B)$$

The formula for conditional probability can be derived from the multiplication rule:

$$P(B|A) = \frac{P(A \cap B)}{P(A)}$$

$$P(A|B) = \frac{P(A \cap B)}{P(B)}$$

The multiplication rule can also be extended for more than two events as follows:

$$P(A \cap B \cap C \cap D) = P(A) \times P(B) \times P(C) \times P(D)$$
$$\text{for independent events } A, B, C, \text{and } D$$

$$P(A \cap B \cap C \cap D) = P(A) \times P(B|A) \times P(C|A \cap B) \times P(D|A \cap B \cap C)$$
$$\text{when events } A, B, C, D \text{ are not independent}$$

Example 3.16

In a target database, there are the following targets:

Category	Priority	Number of targets
Surface-to-air missile (SAM) site	High	12
	Low	24
Sensor site	High	8
	Low	12
Runway	High	3
	Low	7
Facilities	High	1
	Low	5
Total	High	24
	Low	48
	Total	72

a. Two targets are selected randomly from the database without updating the database. What is the probability that one of the selected targets is a SAM site and the other is a facility?

$P(M)$: *The probability that a selected target is a SAM site*

$$P(M) = 36/72 = 0.5$$

$P(F)$: *The probability that a selected target is a facility*

$$P(F) : 6/72 = 0.083$$

$$P(M \cap F) = 0.5 \times 0.083 = 0.042$$

b. Two targets are selected randomly from the database. The database is updated after the first selection (i.e., the selected target is removed from the database). What is the probability that both of the targets are SAM sites?

$$P(M_1) = 0.5 \text{ for the first selection}$$

$$P(M_2) = 35/71 = 0.493$$

$$P(M_1 \cap M_2) = 0.5 \times 0.493 = 0.246$$

c. We select a target randomly, and we know that the selected target is a sensor site. What is the probability that the selected target is a high-priority target? Are selecting a high-priority target H and selecting a sensor site S independent events?

$$P(H) = 24/72 = 0.333$$

$$P(H|S) = 8/20 = 0.4$$

$P(H|S) \neq P(H)$ This indicates that H and S are not independent events.

d. We randomly select a target and we know that the selected target is a SAM site. What is the probability that the selected target is a high-priority target? Are selecting a high-priority target H and selecting a SAM site S independent events?

$$P(H) = 24/72 = 0.333$$

$$P(H|M) = 12/36 = 0.333$$

$P(H|M) \neq P(M)$ This result indicates that H and M are independent events.

3.2.3 Mutually Exclusive Events and Addition Rule

When two events A and B cannot occur together, they are mutually exclusive events. This result is different from being independent. Two independent events can occur at the same time, but the probability for one of them does not change based on the occurrence of the other. However two events A and B are mutually exclusive only if the following are true:

$$P(A|B) = 0 \quad \text{and} \quad P(B|A) = 0$$

For finding the probability that one of two events A or B occurs $P(A \cup B)$, the probability that A occurs $P(A)$ and the probability that B occurs $P(B)$ are summed only if two events are mutually exclusive.

$$P(A \cup B) = P(A) + P(B)$$

This is called the addition rule. When two events A and B are not mutually exclusive, the addition rule becomes as follows:

$$P(A \cup B) = P(A) + P(B) - P(A \cap B)$$

Example 3.17

Use the target database given in Example 3.16.

a. We randomly pick one target from the database. What is the probability that the target is a SAM site or a facility?

$$P(M) = 36/72 = 0.5$$

$$P(F) = 6/72 = 0.083$$

$$P(M \cup F) = 0.5 + 0.083 = 0.583$$

Note that M and F are mutually exclusive, i.e., $P(M|F) = 0$ and $P(F|M) = 0$.

b. We randomly pick one target from the database. What is the probability that the target is a SAM site or a high-priority target?

$$P(M) = 36/72 = 0.5$$

$$P(H) = 24/72 = 0.333$$

$$P(M \cap H) = 0.5 \times 0.333 = 0.167$$

$$P(M \cup H) = 0.5 + 0.333 - 0.167 = 0.667$$

Note that M and H are not mutually exclusive.

Finally, **De Morgan's Law** can also be applied to the probability for changing between multiplication and addition rules as follows:

$$P(\overline{A \cup B}) = P(\overline{A} \cap \overline{B})$$

$$P(\overline{A \cap B}) = P(\overline{A} \cup \overline{B})$$

3.2.4 Total Probability and Bayes' Theorem

The law of total probability is another useful tool for calculating the probability of an event, when we can partition the sample into n partitions and know the conditional probability of the event for each partition. Let us assume that we can divide the sample space into three partitions B_1, B_2, and B_3, such that

$$P(B_1) + P(B_2) + P(B_3) = 1$$

We also know the conditional probability of Event A for each partition (i.e., $P(A|B_1)$, $P(A|B_2)$, $P(A|B_3)$; then the $P(A)$ is given by

$$P(A) = P(A \cap B_1) + P(A \cap B_2) + P(A \cap B_3)$$

Example 3.18

There are three SAMs (i.e., M_1, M_2, M_3) that can engage with an aircraft. One SAM will be selected randomly for the engagement. The probability that one of these SAMs is selected for the engagement is as follows for each SAM:

$P(M_1) = 0.3$
$P(M_2) = 0.2$
$P(M_3) = 0.5$

The probability of hit is different for each SAM from the others and as follows:

$P(A|M_1) = 0.6$
$P(A|M_2) = 0.9$
$P(A|M_3) = 0.3$

What is the total probability of hit $P(A)$?

$$P(A) = P(A \cap M_1) + P(A \cap M_2) + P(A \cap M_3)$$
$$P(A) = P(M_1)P(A|M_1) + P(M_2)P(A|M_2) + P(M_3)P(A|M_3)$$
$$P(A) = 0.3 \times 0.6 + 0.2 \times 0.9 + 0.5 \times 0.3$$
$$P(A) = 0.51$$

In Example 3.18, we are already given the probability of hit for each SAM. Can we reverse those probabilities, such as "What is the probability that an aircraft was engaged with SAM Type 1 (i.e., M_1) knowing that the aircraft was hit?" In other words, what is $P(M_1|A)$?

Let us work on the formulas that we already have:

$$P(A \cap M_1) = P(A|M_1)\, P(M_1)$$
$$P(A \cap M_1) = P(M_1|A)\, P(A)$$
$$P(A|M_1)\, P(M_1) = P(M_1|A)\, P(A)$$

From here we can reach the answer that we are looking for:

$$P(M_1|A) = \frac{P(A|M_1)P(M_1)}{P(A)}$$

When $P(A)$ is replaced with the formula for the total probability, the equation in ***Bayes' Theorem*** for the problem in Example 3.18 is obtained:

$$P(M_1|A) = \frac{P(A|M_1)P(M_1)}{P(M_1)P(A|M_1) + P(M_2)P(A|M_2) + P(M_3)P(A|M_3)}$$

Example 3.19

What is the probability that an aircraft was engaged with SAM Type 1, (i.e., M_1) knowing that the aircraft was hit? Use the data in Example 3.18.

$$P(M_1|A) = \frac{P(A|M_1)P(M_1)}{P(M_1)P(A|M_1) + P(M_2)P(A|M_2) + P(M_3)P(A|M_3)}$$

$$P(M_1|A) = \frac{0.6 \times 0.3}{0.51} = 0.353$$

3.3 RANDOM VARIABLE

A random variable assigns a number to the outcomes of an experiment. For each value that can be used in this process, there is a probability that it is assigned to an outcome. These probabilities create a probability distribution, which can be discrete or continuous. When the assignment is made out of a discrete set of values that can be counted, a discrete probability distribution is needed. If a random variable assigns values from a discrete set of values according to a discrete probability distribution, than it is called a discrete random variable. A continuous random variable assigns values not countable but measurable according to a continuous probability distribution.

In this book, we use capital letters and typically X to indicate a random variable. We use small letters, typically x, as a value that can be assigned by a random variable. In our notation $P(x)$ is the probability that value x is assigned by random variable X:

$$P(x) = P(X = x)$$

3.3.1 Discrete Distributions

A discrete random variable X must assign a value x to an outcome of an experiment according to a discrete probability distribution, which must satisfy the following conditions:

$$0 \leq P(x) \leq 1 \text{ for all } x$$

$$\sum P(x) = 1$$

Example 3.20

a. A random variable X gives the number of times x that a rifleman fires in one minute when the rifleman is in an engagement. The probabilities for each value that can be assigned by X are depicted in the following table. Is this table a valid discrete probability distribution?

X	1	5	10	25
$P(x)$	1	0.5	0.2	0.1

The answer is "NO", because $\sum P(x) = 1.8$, which is bigger than one. Therefore, it does not satisfy the second condition

b. For the same random variable, the following table can be a discrete probability distribution:

X	1	5	10	25
$P(x)$	0.1	0.25	0.5	0.15

In Example 3.20b, there is a set of ordered pairs, which gives the probability for each value that can be assigned by the random variable X. This set of ordered pairs is called a *probability mass function*. Another type of discrete distribution function is called a *cumulative distribution function*, which gives the probability for the values smaller than or equal to a given value x. In this book we use $f(x)$ to indicate the cumulative probability:

$$f(x) = F(C = x) = P(X \leq x)$$

Example 3.21

For the same probability distribution in Example 3.20.b, the following table gives the cumulative probability distribution:

X	1	5	10	25
$f(x)$	0.1	0.35	0.85	1

We can find the expected value $E[X]$, variance $V[X]$, and standard deviation σ for a probability distribution function, which are important to have an insight about the random variable X. $E[X]$, $V[X]$, and σ are equivalent to mean value μ, variance σ^2, and standard deviation σ in descriptive statistics, respectively. The expected value for a random variable X is given by

$$E[X] = \sum_{i=1}^{n} x_i P(x_i)$$

In this equation, n is equal to the number of value-probability pairs in the probability distribution.

Because $E[X] = \mu$, the variance of X can be formulated as follows:

$$V[X] = E[X - \mu]^2$$

$$V[X] = E[X^2 - 2\mu X + \mu^2]$$

$$V[X] = E[X^2] - 2\mu E[X] + E[\mu^2]$$

$$V[X] = E[X^2] - 2\mu^2 + \mu^2$$

$$V[X] = E[X^2] - \mu^2$$

Following this equation, the variance of a probability distribution function can be defined as the difference between the second moment and the square of the first moment of a distribution function and given by

$$V[X] = E[X^2] - E[X]^2$$

Example 3.22

What is the expected value, variance, and standard deviation for the probability distribution in Example 3.20.b:

a. Expected value

$$E[X] = 1 \times 0.1 + 5 \times 0.25 + 10 \times 0.5 + 25 \times 0.15$$

$$\mu = E[X] = 10.1$$

b. Variance

$$E[X^2] = 1 \times 0.1 + 25 \times 0.25 + 100 \times 0.5 + 625 \times 0.15$$

$$E[X^2] = 150.1$$

$$V[X] = E[X^2] - \mu^2$$

$$V[X] = 150.1 - 102.01$$

$$V[X] = 48.09$$

c. Standard deviation

$$\sigma = \sqrt{V[\mathrm{X}]}$$

$$\sigma = 6.935$$

Probability mass functions, expected value, and variance formulations for various discrete probability distributions are depicted in Table 3.1. We explain the essence of each distribution and the conditions for them in the following sections.

3.3.1.1 The Uniform Distribution. A probability distribution defined over n distinct values $x_1, x_2, \ldots, x_n$ equally likely to occur is called uniform distribution:

$$P(X = x_i) = 1/n \quad \text{for} \quad i = 1, 2, \ldots, n$$

In many previous examples in this book, we use the uniform distribution. For example, when a die is rolled, the outcome for rolling a die is uniformly distributed, which means the probability for each possible outcome is the same (i.e., 1/6).

3.3.1.2 The Binomial Distribution. A situation can be described by using binomial distribution when the following conditions are observed:

- A finite number n of trials are carried out,
- Each trial is *identical*, and the probability to observe a success P is the same for each trial,
- Each trial is *independent* from the others.

When these conditions are met, the binomial distribution can be used to answer the following questions:

- What is the probability to observe x number of successes $P(x)$ in n trials?
- What is the expected number of successes $E[X]$ in n trials?

Example 3.23

The probability that a rifleman hits a target when he shoots with his rifle is 0.01. During a battle, the rifleman fires 300 times. ($P = 0.01$, $n = 300$).

a. What is the probability that the rifleman hits two targets during this battle?

$$P(x) = C_x^n P^x q^{n-x}$$

$$P(2) = C_2^{300} \times 0.01^2 \times 0.99^{300-2}$$

$$P(2) = 0.248$$

b. What is the expected number of hits for this rifleman during this battle?

$$E[X] = np$$

$$E[X] = 3$$

3.3.1.3 The Geometric Distribution. The geometric distribution is also for identical and independent events the same as the binomial distribution. Again, for each event, we may have one of two outcomes, and we call one of these outcomes a success. The probability of success for each event is the same and equal to P. Although the conditions are the same, the question that we ask is different from the binomial distribution. Now we would like to learn the probability $P(x)$ that the first success will be at the xth trial.

Example 3.24

For the same conditions in Example 3.22,

a. What is the probability that the rifleman misses in his first 99 trial and hits a target the first time in his 100th trial?

$$P(x) = Pq^{x-1}$$

$$P(2) = 0.01 \times 0.99^{99}$$

$$P(100) = 0.0037$$

b. What is the expected number of times that this rifleman fires before he hits his first target?

$$E[X] = 1/P$$

$$E[X] = 100$$

3.3.1.4 The Negative Binomial Distribution. The negative binomial distribution is almost the same as for the geometric distribution. Instead of the first success, now the question is what is the probability that the xth success occurs in the nth trial is. Note that when x is one, the negative binomial distribution becomes the same as the geometric distribution.

Example 3.25

For the same conditions in Example 3.22,

a. What is the probability that the rifleman hits his second target in his 150th trial?

$$P(x) = C_{x-1}^{n-1} P^x q^{n-x}$$

$$P(2) = 149 \times 0.01^2 \times 0.99^{148}$$

$$P(100) = 0.0034$$

b. What is the expected number of times that this rifleman fires before he hits his second target?

$$E[X] = x/P$$

$$E[X] = 200$$

3.3.1.5 The Hypergeometric Distribution. In the hypergeometric distribution, we know the population size N and the number of times k that a selected event is observed within this population. Then the question is what is the probability that x of randomly selected n observations will be selected.

Example 3.26

A unit has 1 captain, 4 lieutenants, 36 NCOs, and 144 enlisted. During a combat, the unit sustains 15 casualties.

a. Knowing that the casualties are random, that is, the probability of being a casualty is the same for everyone in the unit, what is the probability that two of the lieutenants are among the casualties?

$$P(x) = \frac{C_x^k C_{n-x}^{N-k}}{C_n^N}$$

$$P(2) = \frac{C_2^4 C_{15-2}^{185-4}}{C_{15}^{185}}$$

$$P(2) = 0.0319$$

b. What is the expected number of casualties among the enlisted?

$$E[X] = \frac{nk}{N}$$

$$E[X] = \frac{15 \times 144}{185}$$

$$E[X] = 12$$

3.3.1.6 The Poisson Distribution. The Poisson distribution is another useful distribution, which is often used. As with all the other discrete probability distributions, the Poisson distribution is also related to a continuous distribution. The exponential distribution, which is continuous, is based on the Poisson distribution. We will elaborate on this in the next section.

As with other distributions the in this chapter, the Poisson distribution is also for identical and independent events. Therefore, it is memoryless, which means that the past information does not imply anything for the probability of an event in the future. The probability of success, is always the same regardless of what happened in the previous trial. In the Poisson distribution, typically the rate λ that an event occurs is given, and the probability that a certain number x of events is observed during a given time interval t is asked. Therefore, it is very commonly used in availability models, (i.e., how much equipment goes to maintenance) or traffic generation models.

Example 3.27

An ordnance depot is receiving three combat radios to be fixed every day.

a. What is the expected number of radios that the ordnance depot receives in 5 days?

$$E[X] = \mu = \lambda t$$

$$\mu = 3 \times 5 = 15$$

b. What is the probability that eight radios are received in 5 days?

$$P(x) = \frac{e^{-\mu}\mu^x}{x!}$$

$$P(8) = \frac{e^{-15}15^8}{8!}$$

$$P[8] = 0.0194$$

3.3.2 Continuous Distributions

The difference between discrete and continuous probability distributions can be easily observed in their diagrammatic representations. In Example 3.27, the graphics for several discrete and continuous distributions are depicted (Figure 3.6). Also shown is an example of poisson and exponential distributions (Figure 3.7) A function $y = f(x)$ that gives a continuous distribution is called a probability density function (p.d.f.) and can be represented with a curve as illustrated in the example. However because discrete random distributions are for countable values, their graphics are either histograms or bar graphs. Note again that a function for a discrete random variable is called a probability mass function.

Example 3.28

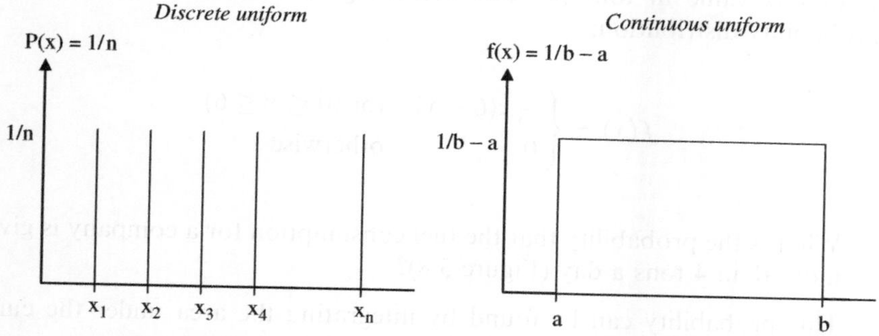

Figure 3.6: Discrete and continuous uniform distributions.

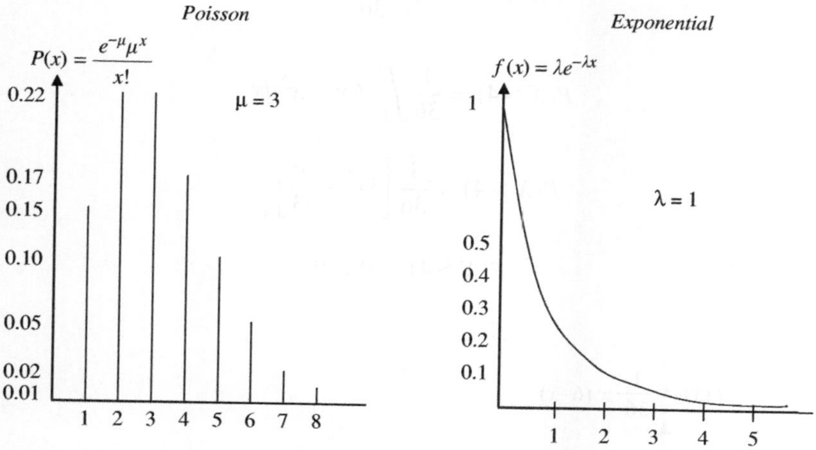

Figure 3.7: Poisson and exponential distributions.

A p.d.f. for a continuous random variable X is specified in a range where x is valid and the function does not return a negative result. When a p.d.f. is available, the probabilities are given by the area under the curve that represents the p.d.f.. When a p.d.f. creates a geometric area, such as the continuous uniform distribution that creates a rectangular region, geometric calculations are enough to find the probabilities. However, generally the curve does not give a geometric shape and therefore integration is needed.

Example 3.29

A continuous random variable X is used to assign a fuel consumption value for mechanized infantry companies in a combat model. Random variable X gives this value in tons per day according to the following continuous probability distribution:

$$f(x) = \begin{cases} \frac{1}{36}x(6-x) & \text{for } (0 \leq x \leq 6) \\ 0 & \text{otherwise} \end{cases}$$

What is the probability that the fuel consumption for a company is given as more than 4 tons a day (Figure 3.8)?

This probability can be found by integrating the area under the curve between 4 and 6:

$$P(X>4) = \int_4^6 \frac{1}{36}x(6-x)dx$$

$$P(X>4) = \frac{1}{36}\int_4^6 6x - x^2 dx$$

$$P(X>4) = \frac{1}{36}\left[3x^2 - \frac{x^3}{3}\right]_4^6$$

$$P(X>4) = 0.259$$

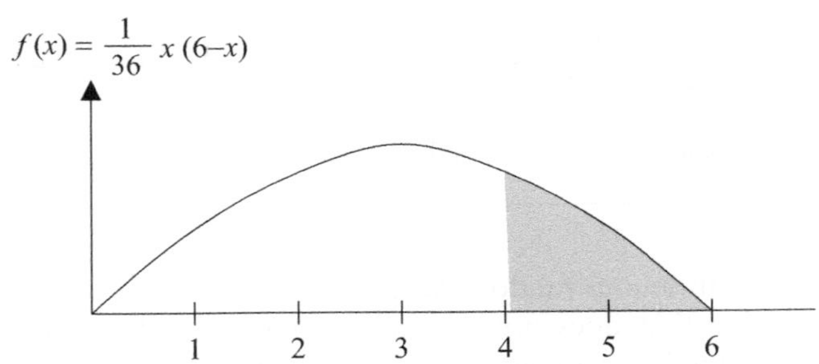

Figure 3.8: Probability distribution for fuel consumption.

The expected value $E[X]$ and variance $V[X]$ of a p.d.f. can also be found through integration:

$$E[X] = \int_a^b xf(x)dx = \mu$$

$$V[X] = \int_a^b x^2 f(x)dx - \mu^2$$

In these equations, a and b are limits of the range where x is valid for the p.d.f. The mean and variance equations together with p.d.f. are shown for various continuous probability distributions in Table 3.4.

3.3.2.1 The Continuous Uniform (Rectangular) Distribution. As in the discrete uniform distribution, the probability is the same for all intervals of the same length in the continuous uniform distribution. It is defined by two parameters a and b, which are the minimum and maximum values that can be assigned to x. The uniform distribution is very useful and often used in combat modeling. For example, mines in a minefield can be scattered randomly according to uniform distribution, which means if there are 10 mines in the field, we presumably find one mine in every 1/10 of the minefield.

Note that the continuous uniform distribution is also called the rectangular distribution because the graph of its density function has a rectangular shape. Other distributions also have a geometric graph shape and, therefore are given geometric names such as triangular distribution.

3.3.2.2 The Exponential Distribution. The exponential distribution is another very useful distribution for simulation and combat modeling. It is often used to model intervals, such mean time between failures, mean time for recovery, and traffic generation for radio networks or transportation networks. In exponential distribution, a constant average rate is given. It is assumed that the number of occurrences during a time interval is distributed with this average according to the Poisson distribution. Then the time between two occurrences is given by the exponential distribution. The reverse of the given rate provides the expected time between two occurrences.

If you examine the distribution closely, it is easy to see that most of the time a random variable using the exponential distribution returns a value smaller than the expected value. When the value returned by the random variable is higher than the expected value, then it is often much higher. Therefore, this distribution is often used to model the things like mean time for recovery. When a failure is a typical one, it can be fixed within the expected time. When the failure is not typical, that may take much longer. Therefore, the exponential distribution can model this realistically.

Example 3.30

A maintenance group at an airbase can service one aircraft a day. What is the probability that the next aircraft will be returned from maintenance in less than one day? Assume that the service rate is distributed according to the Poisson distribution.

$$P(X<1) = \int_0^1 \lambda e^{-\lambda x} dx$$

$$P(X<1) = \left[-e^{-\lambda x}\right]_0^1$$

$$P(X) = 0.632$$

3.3.2.3 The Normal (Gaussian) Distribution. The normal distribution, which is also called the Gaussian distribution, is a continuous distribution that is important in statistics and simulation. The graph of its density function has a bell shape. The highest point in this bell gives the mean value μ, and the width of the bell is determined by variance σ^2. Therefore the normal distribution is characterized by these two parameters μ and σ^2 also called location and scale parameters. When μ is zero and σ^2 is one, the distribution is called the standard normal distribution.

The graph of normal distribution is illustrated for various μ and σ^2 in Figure 3.9. The mean value μ moves the curve to the right or left along the x

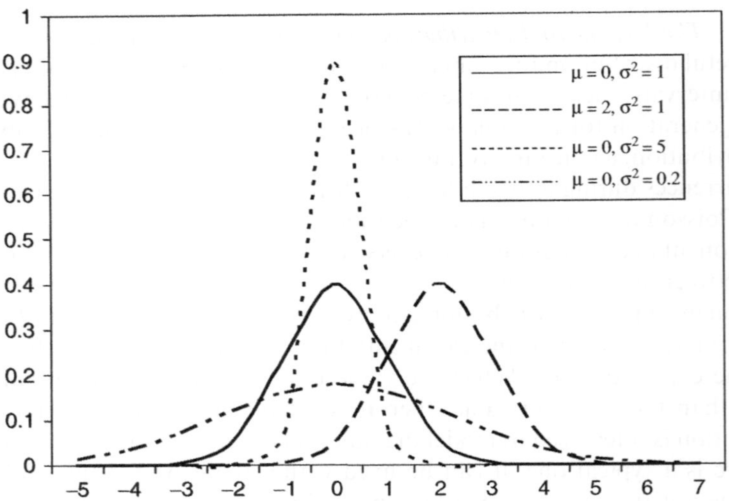

Figure 3.9: The normal distribution.

axis. Therefore, it is called a location parameter. However, variance makes the curve thinner and taller, which indicates less dispersion, or larger and shorter, which indicates more dispersion. Hence when variance σ^2 is larger, the curve gets shorter and larger. Therefore variance σ^2 is called a scale parameter.

One feature of the normal distribution can save us from taking integrations to find the probabilities. The probabilities for the standard normal distribution have been calculated and are provided in Table 3.3. The probabilities for values in the units of standard deviation are the same also for the other normal distributions with different μ and σ. Therefore, when we convert a value x for a random variable X distributed according to any normal distribution into the units of standard deviation, we can look its probability up in the standard normal distribution table. The value in the units of standard deviation is called a z score and is given by

$$z = \frac{x - \mu}{\sigma}$$

Example 3.31

A combat model assigns the depth of submarines according to the uniform distribution with a mean value $\mu = 200$ meters, and a standard deviation $\sigma = 25$ meters, if the CAX operators do not explicitly set this value. What is the probability that the depth of a submarine located by the model is less than 250 meters?

$$z = \frac{x - \mu}{\sigma}$$

$$z = \frac{250 - 200}{25} = 2$$

Go to the row with the z value 2.0 in Table 3.3, and the column 0 to find $P(Z<2)$. Column numbers are for the 1/100 of z scores. For example, if z score was 2.02, we would go to the column marked with 2.

$$P(Z \leq 2) = 0.9772$$

In simulation, the normal distribution is especially important because of the central limit theorem, which states that the sum of many independent and identically distributed random variables is approximately normally distributed. This means that whatever is the number and type of the distributions that we use in a simulation, the final results can be approximated by the normal

Table 3.3: Discrete probability distributions.

Distribution	Parameters	Function $P(x)$	$E[X]$	$V[X]$
Uniform	n: number of distinct values	$p(x_i) = 1/n$	$E[X] = \frac{\sum_{i=1}^{n} x_i}{n}$	$V[X] = \frac{\sum_{i=1}^{n} x_i^2}{n} - \left(\frac{\sum_{i=1}^{n} x_i}{n}\right)^2$
Binomial	n: number of trials P: probability of success $q = 1-p$ x: number of successful trials	$p(x) = C_x^n p^x q^{n-x}$	$E[X] = np$	$V[X] = npq$
Geometric	p: probability of success $q = 1-p$ x: the trial with the first success	$P(x) = Pq^{x-1}$	$E[X] = 1/P$	$V[X] = q/p^2$
Negative Binomial	P: probability of success $q = 1-P$ $x = n$: xth success in nth trial	$P(x) = C_{x-1}^{n-1} P^x q^{n-x}$	$E[X] = x/P$	$V[X] = xq/P^2$
Hypergeometric	N: number of events in a population k: frequency of a given event n: number of events in a sample x: frequency of the event in the given sample	$P(x) = \frac{C_x^k C_{n-x}^{N-k}}{C_n^N}$	$E[X] = \frac{nk}{N}$	$V[X] = \frac{nk}{N}\left(1 - \frac{k}{N}\right)\left(\frac{N-n}{N-1}\right)$
Poisson	λ: the expected rate of occurrence t: the length of a given time interval μ: expected number of occurrence for interval t $\mu = \lambda t$ x: number of occurrence for interval t	$P(x) = \frac{e^{-\mu}\mu^x}{x!}$	$E[X] = \lambda t$	$V[X] = \lambda t$

distribution. Therefore, we can run simulations multiple times, create a sample by the results at the end, and use this sample to build a confidence interval because we know that the results are approximately normally distributed. We will elaborate on the confidence interval more in the next section.

First, let us state the central limit theorem more formally: When the sample size $n \geq 30$, the sampling distribution of the sample mean $\overline{x}$ is approximately normally distributed irrespective of the distribution for the population, and

$$\overline{x} = \mu \quad \text{and} \quad \sigma = \frac{\overline{\sigma}}{\sqrt{n}}$$

3.4 INFERENTIAL STATISTICS

Now we have enough background to start with inferential statistics, which is mainly for two key ideas: estimation and hypothesis testing. In the following section, we focus on estimation, and the later section will be about hypothesis testing. We will also explain another important topic called the goodness of fit test. There are also other important topics in statistics. Which include analysis of variance (ANOVA) and regression. Because this book is not about statistics, this topic are not included here. Readers interested in ANOVA and regression can refer to [CRA02].

3.4.1 Confidence Interval

Sample statistics, such as $\overline{x}$ and $\overline{P}$ can be used to estimate the value of the related population parameters, such as μ and P. A well-known approach for this is that an interval around a point estimate is constructed instead of simply giving the point estimate. This interval around the point estimate is called the confidence interval, and it is constructed from either a z or a t score based on central limit theorem. This interval basically states that a certain percentage of the time the observation in the population will be within the limits of an interval. Because we can approximate the sample mean distribution according to the normal distribution, all we need to find is the values that fall under the curve both at the right and the left of the mean with the total probability that we build our confidence on. If the sample size is (large, i.e., $n > 30$), z scores can be used for this purpose as follows:

$$z = \frac{\overline{x} - \mu}{\overline{\sigma}/\sqrt{n}}$$

$$\mu = \overline{x} \pm z\overline{\sigma}/\sqrt{n}$$

where

$$\overline{\sigma}^2 = \frac{n}{n-1} s^2$$

Note that the probabilities in Table 3.3 are given for all the values smaller than the given z value. Here, we are building the confidence around the mean, which means both tails of the curve will be excluded after the required percentage. In other words, when we are building 95% confidence, we should find the z value to be 0.975 because we will exclude 0.25 from both tails.

Example 3.32

New air-to-ground ammunition is tested against runways. In 100 test firings, the ammunition created a crater with mean value $\overline{x} = 15$ meters and variance $s^2 = 4$ meters. Calculate a 95% confidence interval for the mean crater size in meters created by the new ammunition (Shown in Figure 3.10).

First we need to find the z score for 95%. The tail value for 95% is 2.5% because we need to exclude both tails of the curve. For this we locate the closest value to 0.975 in Table 3.3. The z score for 0.975 is 1.96. We can get the same value also from Table 3.4 for the critical z scores.

$$\mu = \overline{x} \pm z^{\overline{\sigma}}/\sqrt{n}$$

$$\mu = 15 \pm 1.96^{2.01}/\sqrt{100}$$

$$\mu = 15 \pm 0.394$$

Therefore, we are 95% confident that the crater size of this ammunition will be between 14.604 meters and 15.396 meters.

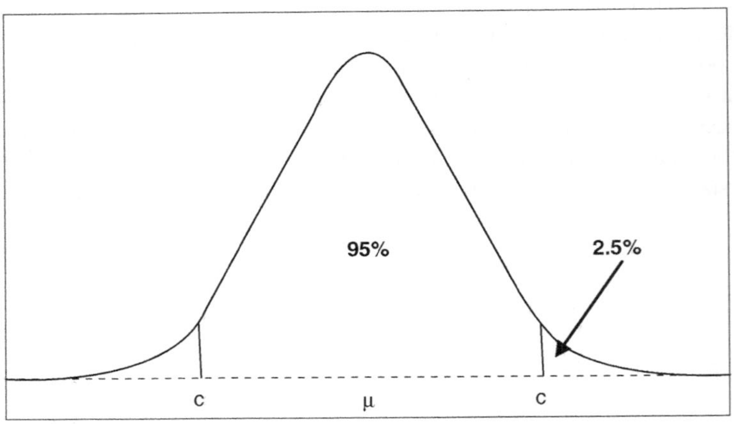

Figure 3.10: Tail value for 95%.

Table 3.4: Continuous probability distributions.

Distribution	Parameters	Function $f(x)$	$E[X]$	$V[X]$
Uniform	a, b: the upper and lower limits of the range	$f(x) = \frac{1}{b-a}$	$E[X] = \frac{a+b}{2}$	$V[X] = \frac{(b-a)^2}{12}$
Exponential	λ: the expected rate of occurrence x: inter occurrence time	$f(x) = \lambda e^{-\lambda x}$	$E[X] = 1/\lambda$	$V[X] = \frac{1}{\lambda^2}$
Normal	μ: mean value σ: variance x: The value for which the probability will be returned	$f(x) = \frac{1}{\sigma\sqrt{2\pi}} e^{-\frac{(x-\mu)^2}{2\sigma^2}}$	$E[X] = \mu$	$V[X] = \sigma^2$

When the sample size is smaller than 30 ($n < 30$), we need to use another distribution called the t-distribution but not the normal distribution. All t-distributions are also symmetric around zero, which is the same as the standard uniform distribution. In the t-distributions, another parameter v called the number of degrees of freedom of the distribution is introduced. In confidence interval calculations, v is equal to $n-1$; therefore it is related to the sample size. As v increases, the t-distribution gets thinner and taller, and it looks more like the standard normal distribution. This is expected because our confidence drops as the sample size decreases. Therefore, we have a larger and shorter bell shape when we have a smaller sample, which a means smaller v. The formulation for the confidence interval calculation does not change much when we use the t-distributions. All we need to do is replace the z with the t score, which can be looked up in Table 3.5.

Example 3.33

For the same problem in Example 3.31, calculate the 95% confidence interval when the sample size is 20.

First we need to find the t score for 95%. The degrees of freedom is $v = 20 - 1 = 19$. Therefore, we go row 19 in Table 3.5. Because we are looking for 95% confidence, the tail value is 0.025, and therefore, we go to the column marked as 0.975. The t score for $P = 0.975$ and $v = 19$ is 2.093.

$$\mu = \overline{x} \pm t^{\overline{\sigma}}/\sqrt{n}$$

$$\mu = 15 \pm 2.093^{2.02}/\sqrt{100}$$

$$\mu = 15 \pm 0.423$$

Therefore, we are 95% confident that the crater size of this ammunition will be between 14.577 meters and 15.423 meters.

3.4.2 Hypothesis Test

This section is dedicated to the other important topic in inferential statistics called the hypothesis test, which is actually an extension of the ideas from the previous section with z and t scores. In hypothesis testing, we will be given a mean or probability value, which is believed to be true or false, and we will try to prove or disprove that. We will go through the stages of a hypothesis test one by one working on the same example:

Stage 1: State the variable being considered.
In this stage, the variable that will be tested is stated clearly.

Table 3.5: Probabilities for the standard normal distribution.

z	0	1	2	3	4	5	6	7	8	9
0.0	0.5000	0.5040	0.5080	0.5120	0.5160	0.5199	0.5239	0.5279	0.5319	0.5359
0.1	0.5398	0.5438	0.5478	0.5517	0.5557	0.5596	0.5636	0.5675	0.5714	0.5753
0.2	0.5795	0.5832	0.5871	0.5910	0.5948	0.5987	0.6026	0.6064	0.6103	0.6141
0.3	0.6179	0.6217	0.6255	0.6293	0.6331	0.6368	0.6406	0.6443	0.6480	0.6517
0.4	0.6554	0.6591	0.6628	0.6664	0.6700	0.6736	0.6772	0.6808	0.6844	6879
0.5	0.6915	0.6950	0.6985	0.7019	0.7054	0.7088	0.7123	0.7157	0.7190	0.7224
0.6	0.7257	0.7291	0.7324	0.7357	0.7389	0.7422	0.7454	0.7486	0.7157	0.7549
0.7	0.7580	0.7611	0.7642	0.7673	0.7704	0.7734	0.7764	0.7794	0.7823	0.7852
0.8	0.7881	0.7910	0.7939	0.7969	0.7995	0.8023	0.8051	0.8078	0.8106	0.8133
0.9	0.8159	0.8186	0.8212	0.8238	0.8264	0.8289	0.8315	0.8340	0.8365	0.8389
1.0	0.8413	0.8438	0.8461	0.8485	0.8508	0.8513	0.8554	0.8577	0.8529	0.8621
1.1	0.8643	0.8665	0.8686	0.8708	0.8729	0.8749	0.8770	0.8790	0.8810	0.8830
1.2	0.8849	0.8869	0.8888	0.8907	0.8925	0.8944	0.8962	0.8980	0.8997	0.9015
1.3	0.9032	0.9049	0.9066	0.9082	0.9099	0.9115	0.9131	0.9147	0.9162	0.9177
1.4	0.9192	0.9207	0.9222	0.9236	0.9215	0.9265	0.9279	0.9292	0.9306	0.9319
1.5	0.9332	0.9345	0.9357	0.9370	0.9382	0.9394	0.9406	0.9418	0.9492	0.9441
1.6	0.9452	0.9463	0.9474	0.9484	0.9495	0.9505	0.9515	0.9525	0.9535	0.9545
1.7	0.9554	0.9564	0.9573	0.9582	0.9591	0.9599	0.9608	0.9616	0.9625	0.9633
1.8	0.9641	0.9649	0.9656	0.9664	0.9671	0.9678	0.9686	0.9693	0.9699	0.9706
1.9	0.9713	0.9719	0.9726	0.9732	0.9738	0.9744	0.9750	0.9756	0.9761	0.9767
2.0	0.9772	0.9778	0.9783	0.9788	0.9793	0.9798	0.9803	0.9808	0.9812	0.9817
2.1	0.9821	0.9826	0.9830	0.9834	0.9838	0.9842	0.9846	0.9850	0.9854	0.9857
2.2	0.9861	0.9864	0.9868	0.9871	0.9875	0.9878	0.9881	0.9884	0.9887	0.9890
2.3	0.9893	0.9896	0.9898	0.9901	0.9904	0.9906	0.9909	0.9911	0.9913	0.9916
2.4	0.9918	0.9920	0.9922	0.9925	0.9927	0.9929	0.9931	0.9932	0.9934	0.9936
2.5	0.9938	0.9940	0.9941	0.9943	0.9945	0.9946	0.9948	0.9949	0.9951	0.9952
2.6	0.9953	0.9955	0.9956	0.9957	0.9959	0.9960	0.9961	0.9962	0.9963	0.9964
2.7	0.9965	0.9966	0.9967	0.9968	0.9969	0.9970	0.9971	0.9972	0.9973	0.9974
2.8	0.9974	0.9975	0.9976	0.9977	0.9977	0.9978	0.9979	0.9979	0.9980	0.9981
2.9	0.9981	0.9982	0.9982	0.9983	0.9984	0.9984	0.9985	0.9985	0.9986	0.9986
3.0	0.9987	0.9987	0.9987	0.9988	0.9988	0.9989	0.9989	0.9989	0.9990	0.9990
3.1	0.9990	0.9991	0.9991	0.9991	0.9992	0.9992	0.9992	0.9992	0.9993	0.9993
3.2	0.9993	0.9993	0.9994	0.9994	0.9994	0.9994	0.9994	0.9995	0.9995	0.9995
3.3	0.9995	0.9995	0.9995	0.9996	0.9996	0.9996	0.9996	0.9996	0.9996	0.9997
3.4	0.9997	0.9997	0.9997	0.9997	0.9997	0.9997	0.9997	0.9997	0.9997	0.9998

If Z has a normal distribution with $\mu = 0$ and $s^2 = 1$, (i.e., the standard normal distribution), the table gives $P(Z \leq z)>$ for each value of z. For negative values of z, use $1-P(Z \leq z)$.

Example 3.34.a

The U.S. Army expects the average physical training score $\mu = 250$ with standard deviation $\sigma = 5$ for the infantry units. An inspection team randomly selects $n = 50$ people from the 1st Infantry Batallion and tests them. The average for the tested personnel is found as $\overline{x} = 247$. Did the 1st

Infantry Battalion pass the physical training inspection when the result should be significant at the $\alpha = 5\%$ level?

Stage 2: State the ***null and alternative hypothesis***, (i.e., H_o and H_a, respectively).

The null hypothesis H_o is something what we would like to prove. If we are testing mean value μ, it is written as follows:

$$H_o\colon \mu = x$$

Alternative hypothesis H_a is stated based on our suspect. We may suspect that the actual value is less than, more than, or different from the value in the null hypothesis, which is written as below, respectively:

$H_a\colon \mu < x$
$H_a\colon \mu > x$
$H_a\colon \mu \neq x$

Depending on these three conditions, (i.e., less than, more than, or different from), we applied lower tail, upper tail or two tailed tests respectively. Our example is for an upper tail test, which means the critical region is in the upper tail of the normal distribution curve. We will clarify this when we are explaining, at Stage 3.

Example 3.34.b

For Example 3.34.a:

$Ho\colon \mu = 250$
$H_a\colon \mu < 250$ (lower tail test)

Stage 3: Select the distribution to use and determine the rejection criterion.
Based on the sample size, we decide to use the standard normal distribution or a t-distribution. Then we find the rejection criterion according to that distribution. To be able to determine the rejection criterion, a significance level α should also be given. The critical value, which is the border of the critical region, is determined according to the significance level α and the alternative hypothesis H_a. The significance level indicates which percentage of the area under the curve is within the critical region. In other words, it indicates how much the test value $\overline{x}$ should be away from the mean μ to reject the null hypothesis H_o. An alternative hypothesis indicates in which tail the critical region is locates. For example, if it is an upper tail test, (i.e., $H_a\colon \mu > x$), we reject the null hypothesis H_o when $\overline{x} > c$. Alternatively, we can also use z scores to indicate the critical value.

Example 3.34.c

Because the significance level is given as 0.05, we need to find the critical value for $P = 0.95$ such that $P(Z \leqslant z) = P$. We can look it up in Table 3.4. The result for $P = 0.95$ is 1.645. This would be fine if we are doing an upper tail test. Because this is a lower tail test, the critical value is –1.645 (shown in Figure 3.11).

Stage 4: Calculate the value of the test statistics and decide whether it is in the critical region.

Finally we calculate the value for the test statistic and compare the result against the critical value. We already know how to calculate z value for a test statistic:

$$z = \frac{\overline{x} - \mu}{\sigma/\sqrt{n}}$$

Example 3.34.d

$$z = \frac{\overline{x} - \mu}{\sigma/\sqrt{n}}$$

$$z = \frac{247 - 250}{5/\sqrt{50}} = -4.243$$

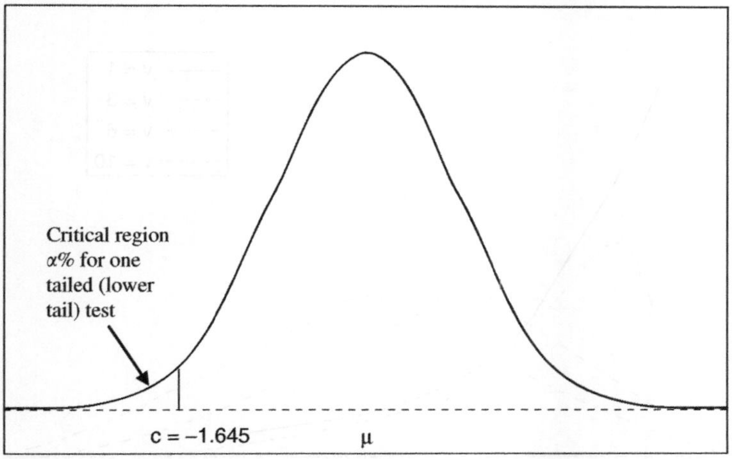

Figure 3.11: Critical region for lower tail test.

Because $-4.243 < -1.645$, the test statistic, (i.e., the average results from the inspection) is in the critical region, and therefore, we reject the null hypothesis, which means the 1st Batalian failed the physical training inspection.

3.4.3 Goodness of Fit

The goodness of fit tests mainly for modeling data, and therefore, it is an important tool set especially for simulation. The descriptive statistics, such as mean and variance, are not enough to model the practical data. We also need to know the distribution of the data. When we have them both, then we can generate random input for our models in a simulation.

One of the methods for the goodness-of-fit test is called the χ^2 test, which we will focus on in this section. The χ^2 goodness-of-fit test is in essence a hypothesis test, where the null hypothesis H_o is that the selected distribution models the data; the alternative hypothesis H_a is that it does not. The first important point that we need to know is that a distribution called χ^2 is used for the χ^2 goodness-of-fit test. The shape of the χ^2 distribution is based on one parameter, degrees of freedom v, as illustrated in Figure 3.12. When the degrees of freedom is less than or equal to 2, the shape of the χ^2 distribution is single tailed. As the degrees of freedom increases, the distribution becomes first two tailed and later less skewed. The degrees of freedom is the number of independent variables used in calculating the test statistic.

Because the approach taken in the χ^2 goodness-of-fit test is the same as the hypothesis test, we do not explain it but give an example on it. Note that the χ^2

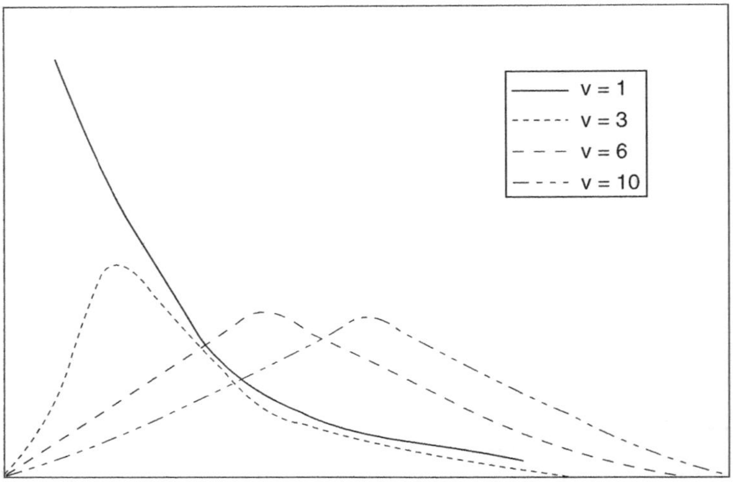

Figure 3.12: The χ^2 distribution.

Table 3.6: Critical values for the standard normal distribution.

p	0.75	0.90	0.95	0.975	0.99	0.995	0.9975	0.999	0.9995
z	0.674	1.282	1.645	1.960	2.326	2.576	2.807	3.090	3.291

The table gives z value such that $P(Z \leq z) = P$ where Z has the standard normal distribution.

goodness-of-fit test is conducted as an upper tail test, and the critical values for the χ^2 distribution are given in Table 3.6.

Example 3.35

A data model for the number of aircrafts in maintenance at a time in a given base during an operation is needed. For this reason, the data from one of the latest campaigns are available where the numbers of aircrafts in maintenance every day for 50 consecutive days are given. The frequency table for these data is as follows:

# of aircrafts in maintenance	**2**	**3**	**4**	**5**	**6**	**Total**
frequency	8	11	9	10	12	50

Stage 1: State the variable being considered.

It is already stated in the problem explanation: the distribution for the number of aircrafts in maintenance in the base.

Stage 2: State the null and alternative hypothesis. Table 3.7 and 3.8

The data seem to be distributed randomly according to uniform distribution between 2 and 6. Still it is good to start with plotting the data and to examine the shape of the plot before stating the hypothesis about the distribution (shown in Figure 3.13):

Although the frequency for 6 and 3 is higher, and for 2 is lower, it still looks like the data are randomly distributed between 2 and 6. Note that this is a discrete but not a continuous distribution. Please also note that when it is a continuous distribution, the procedure for the test does not change.

We are now ready to state our hypothesis:

H_0: The number of aircrafts in maintenance is distributed according to the discrete uniform distribution.

H_a: It is NOT distributed according to the discrete uniform distribution.

Table 3.7: Critical values for the t-distribution.

	P = 0.75	0.80	0.85	0.90	0.95	0.975	0.99	0.995	0.9975	0.9990	0.9995
V = 1	1.000	1.376	1.963	3.078	6.314	12.71	31.82	63.66	127.3	318.3	636.6
2	0.816	1.061	1.386	1.886	2.920	4.303	6.965	9.925	14.09	22.33	31.60
3	0.765	0.978	1.250	1.638	2.353	3.182	4.541	5.841	7.453	10.21	12.92
4	0.741	0.941	1.190	1.533	2.132	2.776	3.747	4.604	5.598	7.173	8.610
5	0.727	0.920	1.156	1.476	2.015	2.571	3.365	4.032	4.773	5.893	6.869
6	0.718	0.906	1.134	1.440	1.943	2.447	3.143	3.707	4.317	5.208	5.959
7	0.711	0.896	1.119	1.415	1.895	2.365	2.998	3.499	4.029	4.785	5.408
8	0.706	0.889	1.108	1.397	1.860	2.306	2.896	3.355	3.833	4.501	5.041
9	0.703	0.883	1.100	1.383	1.833	2.262	2.821	3.250	3.690	4.297	4.781
10	0.700	0.879	1.093	1.372	1.812	2.228	2.764	3.169	3.581	4.144	4.587
11	0.697	0.876	1.088	1.363	1.796	2.201	2.718	3.106	3.497	4.025	4.437
12	0.695	0.873	1.083	1.356	1.782	2.179	2.681	3.055	3.428	3.930	4.318
13	0.694	0.870	1.079	1.350	1.771	2.160	2.650	3.012	3.372	3.852	4.221
14	0.692	0.868	1.076	1.345	1.761	2.145	2.624	2.977	3.326	3.787	4.140
15	0.691	0.866	1.074	1.341	1.753	2.131	2.602	2.947	3.286	3.733	4.073
16	0.690	0.865	1.071	1.337	1.746	2.120	2.583	2.921	3.252	3.686	4.015
17	0.689	0.863	1.069	1.333	1.740	2.110	2.567	2.898	3.222	3.646	3.965
18	0.688	0.862	1.067	1.330	1.734	2.101	2.552	2.878	3.197	3.610	3.922
19	0.688	0.861	1.066	1.328	1.729	2.093	2.539	2.861	3.174	3.579	3.883
20	0.687	0.860	1.064	1.325	1.725	2.086	2.528	2.845	3.153	3.552	3.850
21	0.686	0.859	1.063	1.323	1.721	2.080	2.518	2.831	3.135	3.527	3.819
22	0.686	0.858	1.061	1.321	1.717	2.074	2.508	2.819	3.119	3.505	3.792
23	0.685	0.858	1.060	1.319	1.714	2.069	2.500	2.807	3.104	3.485	3.767
24	0.685	0.857	1.059	1.318	1.711	2.064	2.492	2.797	3.091	3.467	3.745
25	0.684	0.856	1.058	1.316	1.708	2.060	2.485	2.787	3.078	3.450	3.725
26	0.684	0.856	1.058	1.315	1.706	2.056	2.479	2.779	3.067	3.435	3.707
27	0.684	0.855	1.057	1.314	1.703	2.052	2.473	2.771	3.057	3.421	3.690
28	0.683	0.855	1.056	1.313	1.701	2.048	2.467	2.763	3.047	3.408	3.674
29	0.683	0.854	1.055	1.311	1.699	2.045	2.462	2.756	3.038	3.396	3.659
30	0.683	0.854	1.055	1.310	1.697	2.042	2.457	2.750	3.030	3.385	3.646
40	0.681	0.851	1.050	1.303	1.684	2.021	2.423	2.704	2.971	3.307	3.551
50	0.679	0.849	1.047	1.299	1.676	2.009	2.403	2.678	2.937	3.261	3.496
60	0.679	0.848	1.045	1.296	1.671	2.000	2.390	2.660	2.915	3.232	3.460
80	0.678	0.846	1.043	1.292	1.664	1.990	2.374	2.639	2.887	3.195	3.416
100	0.677	0.845	1.042	1.290	1.660	1.984	2.364	2.626	2.871	3.174	3.390
120	0.677	0.845	1.041	1.289	1.658	1.980	2.358	2.617	2.860	3.160	3.373
∞	0.674	0.842	1.036	1.282	1.645	1.960	2.326	2.576	2.807	3.090	3.291

The table gives the value of t such that $P(T \leq t) = P$ for each pair of values, the degree of freedom v and probability P.

Stage 3: Determine the rejection criterion

The first step in this stage is to find the degrees of freedom v given by

$$v = \text{number of classes} - \text{number of restrictions}$$

Table 3.8: Critical values for the χ^2 distribution.

	P = 0.995	0.975	0.20	0.10	0.05	0.025	0.02	0.01	0.005	0.002	0.001
V = 1	0.0000393	0.000982	1.642	2.706	3.841	5.024	5.412	6.635	7.879	9.550	10.828
2	0.0100	0.0506	3.219	4.605	5.991	7.378	7.824	9.210	10.597	12.429	13.816
3	0.0717	0.216	4.642	6.251	7.815	9.348	9.837	11.345	12.838	14.796	16.266
4	0.207	0.484	5.989	7.779	9.488	11.143	11.668	13.277	14.860	16.924	18.467
5	0.412	0.831	7.289	9.236	11.070	12.833	13.388	15.086	16.750	18.907	20.515
6	0.676	1.237	8.558	10.645	12.592	14.449	15.033	16.812	18.548	20.791	22.458
7	0.989	1.690	9.803	12.017	14.067	16.013	16.622	18.475	20.278	22.601	24.322
8	1.344	2.180	11.030	13.362	15.507	17.535	18.168	20.090	21.955	24.352	26.124
9	1.735	2.700	12.242	14.684	16.919	19.023	19.679	21.666	23.589	26.056	27.877
10	2.156	3.247	13.442	15.987	18.307	20.483	21.161	23.209	25.188	27.722	29.588
11	2.603	3.816	14.631	17.275	19.675	21.920	22.618	24.725	26.757	29.354	31.264
12	3.074	4.404	15.812	18.549	21.026	23.337	24.054	26.217	28.300	30.957	32.909
13	3.565	5.009	16.985	19.812	22.362	24.736	25.472	27.688	29.819	32.535	34.528
14	4.075	5.629	18.151	21.064	23.685	26.119	26.873	29.141	31.319	34.091	36.123
15	4.601	6.262	19.311	22.307	24.996	27.488	28.259	30.578	32.801	35.628	37.697
16	5.142	6.908	20.465	23.542	26.296	28.845	29.633	32.000	34.267	37.146	39.252
17	5.697	7.564	21.615	24.769	27.587	30.191	30.995	33.409	35.718	38.648	40.790
18	6.265	8.231	22.760	25.989	28.869	31.526	32.346	34.805	37.156	40.136	42.312
19	6.844	8.907	23.900	27.204	30.144	32.852	33.687	36.191	38.582	41.610	43.820
20	7.434	9.591	25.038	28.412	31.410	34.170	35.020	37.566	39.997	43.072	45.315
21	8.034	10.283	26.171	29.615	32.671	35.479	36.343	38.932	41.401	44.522	46.797
22	8.643	10.982	27.301	30.813	33.924	36.781	37.659	40.289	42.796	45.962	48.268
23	9.260	11.689	28.429	32.007	35.172	38.076	38.968	41.638	44.181	47.391	49.728
24	9.886	12.401	29.553	33.196	36.415	39.364	40.270	42.980	45.559	48.812	51.179
25	10.520	13.120	30.675	34.382	37.652	40.646	41.566	44.314	46.928	50.223	52.620
26	11.160	13.844	31.795	35.563	38.885	41.923	42.856	45.642	48.290	51.627	54.052
27	11.808	14.573	32.912	36.741	40.113	43.195	44.140	46.963	49.645	53.023	55.476
28	12.461	15.308	34.027	37.916	41.337	44.461	45.419	48.278	50.993	54.411	56.892
29	13.121	16.047	35.139	39.087	42.557	45.722	46.693	49.588	52.336	55.792	58.301
30	13.787	16.791	36.250	40.256	43.773	46.979	47.962	50.892	53.672	57.167	59.703
31	14.458	17.539	37.359	41.422	44.985	48.232	49.226	52.191	55.003	58.536	61.098
32	15.134	18.291	38.466	42.585	46.194	49.480	50.487	53.486	56.328	59.899	62.487
33	15.815	19.047	39.572	43.745	47.400	50.725	51.743	54.776	57.648	61.256	63.870
34	16.501	19.806	40.676	44.903	48.602	51.966	52.995	56.061	58.964	62.608	65.247
35	17.192	20.569	41.778	46.059	49.802	53.203	54.244	57.342	60.275	63.955	66.619
36	17.887	21.336	42.879	47.212	50.998	54.437	55.489	58.619	61.581	65.296	67.985
37	18.586	22.106	43.978	48.363	52.192	55.668	56.730	59.893	62.883	66.633	69.346
38	19.289	22.878	45.076	49.513	53.384	56.896	57.969	61.162	64.181	67.966	70.703
39	19.996	23.654	46.173	50.660	54.572	58.120	59.204	62.428	65.476	69.294	72.055
40	20.707	24.433	47.269	51.805	55.758	59.342	60.436	63.691	66.766	70.618	73.402
41	21.421	25.215	48.363	52.949	56.942	60.561	61.665	64.950	68.053	71.938	74.745
42	22.138	25.999	49.456	54.090	58.124	61.777	62.892	66.206	69.336	73.254	76.084
43	22.859	26.785	50.548	55.230	59.304	62.990	64.116	67.459	70.616	74.566	77.419
44	23.584	27.575	51.639	56.369	60.481	64.201	65.337	68.710	71.893	75.874	78.750
45	24.311	28.366	52.729	57.505	61.656	65.410	66.555	69.957	73.166	77.179	80.077
46	25.041	29.160	53.818	58.641	62.830	66.617	67.771	71.201	74.437	78.481	81.400
47	25.775	29.956	54.906	59.774	64.001	67.821	68.985	72.443	75.704	79.780	82.720
48	26.511	30.755	55.993	60.907	65.171	69.023	70.197	73.683	76.969	81.075	84.037
49	27.249	31.555	57.079	62.038	66.339	70.222	71.406	74.919	78.231	82.367	85.351
50	27.991	32.357	58.164	63.167	67.505	71.420	72.613	76.154	79.490	83.657	86.661

The table gives the value of x such that $P(X \geq x) = P$ for each pair of values, the degree of freedom v and probability P.

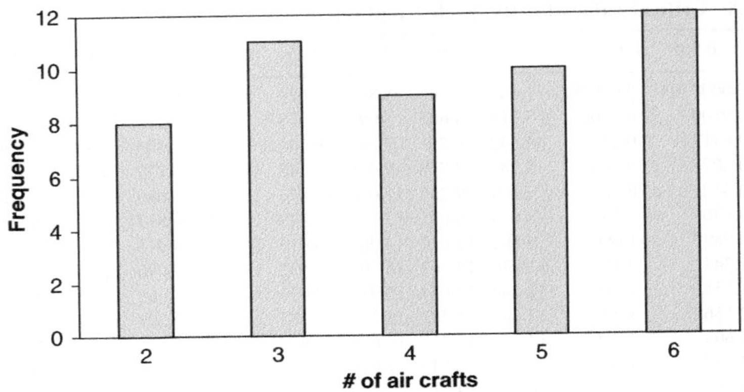

Figure 3.13: Number of aircrafts in maintenance.

In our data set, we have five classes of data and only one restriction, which is that the total frequency is 50. Therefore,

$$v = 5 - 1 = 4$$

The next thing that we need to know is the significance level α, which is dependent on the person making the test or asking the test as we already explained. Let us pick a typical significance level

$$\alpha = 5\%$$

Now we need to look up the critical value for $\chi^2_{5\%}(4)$ up in Table 3.6.

$$\chi^2 5\%(4) = 9.488$$

Stage 4: Calculate the value for the test statistic and compare it with to the rejection criteria

The χ^2 test compares the observed frequency O with the expected frequency E, and a decision is made based on the results of this comparison. Therefore, in this final stage, the difference between the frequencies from the test data for each data class and their expected frequencies are compared. Because we are testing for the discrete uniform distribution, the expected frequency value is

$$E = \frac{\sum_{i=1}^{n} x_i}{n} = \frac{50}{5} = 10$$

We use the following formula to determine the differences between the test data frequencies and the expected frequencies:

$$X^2 = \sum_{i=1}^{n} \frac{(O - E)^2}{E}$$

$$X^2 = 0.4 + 0.1 + 0.1 + 0 + 0.4 = 1.0$$

Because $1.0 < 9.488$, we cannot reject the null hypothesis, and we can assume that the observed data are distributed randomly according to the discrete uniform distribution. This conclusion is significant at the 5% level.

3.5 REVIEW QUESTIONS

3.1 What is a sample, a population, a sample statistic, and a population parameter?

3.2 Why do we need a sample? Discuss the ways to take a sample from a population?

3.3 What are the measures for central tendency?

3.4 What are the measures for dispersion?

3.5 For the following data array 35, 7, 11, 23, 27, 11, 15, 19, 23, 23, 27, 15, 31, 19, 19, 19, 15, 15, 11, 19, 23, 3, 27, 19, 19, 15, 23

a. Draw a frequency polygon, pie chart, and frequency histogram.
b. Is this a skewed distribution?
c. Calculate mean, median, and mode. Explain the relation between these statistics. Does this relation imply anything? Why?
d. Calculate range, variance, and deviation. Does normal distribution imply anything for the measures of dispersion? Why?

3.6 There are four targets and a shooter. The shooter selects one of the targets randomly. The shooter may select a target again even if it was shot and hit before. The probability of a hit is 0.6. The shooter shoots six times

a. What is the probability that Target X is shot at least one time?
b. What is the probability that Target X is hit at least one time?
c. What is the probability that the shooter hits a target all six times?
d. What is the probability that the shooter misses everytime?
e. What is the probability that the shooter hits two times?
f. What is the probability that the shooter hits a target the first time at his third trial?
g. What is the probability that Target X is never shot?
h. What is the probability that the shooter selects Target X all six times?
i. What is the probability that Target X is hit six times?

3.7 What is the difference between two mutually exclusive and two independent events?

3.8 Explain the parameters for the Uniform, Binomial, Geometric, Negative Binomial, Hypergeometric, and Poisson distributions.

3.9 What is the difference between the discrete and continuous uniform distributions? Give an example for each of them to explain the difference.

3.10 For Example 3.29,

a. What is the probability that the fuel consumption for the company is given less than 2 tons per day?

b. What is the probability that the fuel consumption for the company is given between 2 and 4 tons per day?

3.11 The maintenance team can service two aircrafts a day. What is the probability that the next aircraft will be returned from maintenance within 12 hours?

3.12 Twenty soldiers are selected randomly. Their run 2400 meters. The mean value for their results $\overline{x} = 8.85$ minutes and variance $s^2 = 30$ seconds

a. Calculate a 95% confidence interval for the mean basic fitness test result for the battalion.

b. Prove or disprove that 8 out of 10 randomly selected soldiers can run 2400 meters in less than 8.6 minutes.

4

SIMULATION

Simulation is the imitation of a facility, phenomenon, or process often called a system to gain insight about it. A system is a collection of components that interact and act together to achieve an objective. A physical or an abstract representation of the system under study constitutes the model of the system. A model is generally a simplified version of the system, and the simplification is often achieved by making assumptions. A system can be represented also as a collection of many interacting complex models. If the model or the collection of models is simple enough, an analytical approach, such as calculus, algebra, probability, statistics, and so on, can be used to study the system. When the model is too complex for an analytical approach, simulation becomes the tool for studying.

Simulation can be very useful for various purposes, including the following:

- ***Simulation-based acquisition***: Alternative systems can be tried against the requirements, and the most effective system best fit for the requirements among the alternatives can be selected. Alternatively, the criteria that best fits the requirements can be determined through simulation, and the most cost-effective system that satisfies the criteria can be selected.
- ***Analysis of plans***: Future plans can be tried under various assumptions by using simulation, and an insight about the details related to the implementation of the plan can be gained.
- ***Evaluation of a new system, protocol, or tool***: New systems, protocols, and tools can be tried under assumptions, and their performance can be

Computer Assisted Exercises and Training: A Reference Guide, by Erdal Çayirci and Dusan Marincic

evaluated. Hence, the decisions to implement them can be taken more safely.

- ***Training***: Personnel can be trained on the operation of systems and procedures by using simulation.

As we explained in the Introduction, simulation systems can be categorized as live, virtual, and constructive simulations. Let us examine these categories one more time from a different perspective:

Live simulations are the simulations in which real people (i.e., operators of a system) and the real system are used. Often, the interactions of the system with the external environment or the interactions of the components of the system are simulated. For example, a real car is driven in a closed environment where various road and weather conditions are created, and the performance of the car is observed.

In virtual simulations, systems are modeled as realistic as possible either scaled or at the same scale physically or by using computer-generated images and effects. A physical model of the car used in crash tests, a physical model of an airplane in a wind tunnel, and an aircraft simulator fall into this category. In virtual simulations, the operators using the systems, as in the example of an aircraft simulator, are the real people. Virtual simulations can be classified as physical, interactive, and human-in-the-loop simulations. In physical simulations, a more simple, often scaled, physical imitation of a real system is used. Interactive and human-in-the-loop simulation are the same class and mean a system where users interact with synthetic environments, such as simulators.

Finally, in constructive simulations, everything is simulated often by a computer, and the simulation systems are constructed by bringing many abstract models together. A meteorological simulator used for weather forecasts and a communications simulator used for examining the performance of a new communications protocol set are examples of this category. This category is also often called computer simulation.

There are also other approaches in categorizing the simulation systems. We can classify a simulation as follows:

- Deterministic or stochastic
- Static or dynamic

If a simulation does not contain any randomness, it is called deterministic. These types of simulations can be used for complex systems with deterministic states and state changes when a certain set of parameters are passed. A deterministic simulation system can take a long time to run in a computer and can simulate systems like a chemical process, where the outcome can be calculated deterministically based on the inputs and interactions. Stochastic simulations are the simulations in which random variables are required. Almost all of the military constructive simulation systems are stochastic simulations.

A static simulation is the representation of the system at a certain time. It gives the status of a system at a time based on the previous state of the system and given parameters. Static simulation systems can be deterministic or stochastic. Dynamic simulations are run for a time interval. They can be discrete or continuous. In discrete simulations, simulation is forwarded from one point to the next point, and at each point, all state variables are updated. In continuous simulation, the state variables are updated continuously.

4.1 PSEUDORANDOM NUMBER GENERATION AND REALIZATION OF RANDOM VARIABLES

As stated, most constructive military simulations are stochastic simulations; here, stochastic means randomness provided by pseudorandom number generators in a simulation. Understanding pseudorandom number generators is important for a simulation user. For example, if you do not change random number seeds and run the simulation with the same inputs every time, the results will be the same for each run. This result is not expected for stochastic simulations. When a simulation is run multiple times for building a confidence interval, the random number seeds must be changed in every run. Similarly, if we use the same instance of a random number generator for multiple random variables, the distribution of the random values assigned by the random variables may be different from what is expected. Therefore, different instances of the random number generator should be run for different random variables to obtain the random numbers according to the intended distribution. The aim of this section is not only to explain how to generate random numbers but also to clarify key issues like the ones above, and to provide the users with a better understanding of randomness.

4.1.1 Pseudorandom Number Generation

Random number generators compose numbers uniformly distributed between zero and one $U(0, 1)$. Although they are called random numbers, they are not generated randomly. They are actually the outcome of a deterministic function. However, they show the characteristics of random numbers distributed according to the uniform distribution, and this is true for any number of consecutive random numbers generated. For example, if you pick any number of consecutive numbers generated by a random number generator and run the χ^2 goodness-of-fit test on them, you should see that they are randomly distributed. The bottom line is that although the set of numbers obtained by these generators is randomly distributed, the process that generates those numbers is not random. Therefore, we call this process pseudorandom number generation.

Example 4.1

The Mid Square Method

Step 1: Select a four-digit positive integer as a **random number seed.**

$$Zo = 1421$$

Step 2: Square the previous four-digit number, which is the seed at the beginning. If the obtained number is not eight digits, then add zeros to the left of the number until it becomes eight digits.

02019241

Step 3: Take the middle four digits as your next four-digit number. Place a decimal point at the left of this four-digit number to obtain the random number. If you need more random numbers, input the new four-digit number into Step 2, and continue.

$$Z_1 = 0192$$

Random number is 0.0192

Random numbers generated by the mid square method starting from the seed 1421:

0192, 0368, 1354, 8333, 4388, 2545, 4770, 7529, 6858, 0321, 1030, 0609, 3708, 7492, 1300, 6900, 6100, 2100, 4100, 8100, 6100.

Random numbers generated by the mid square method starting from the seed 1009:

0180, 0324, 1049, 1004, 0080, 0064, 0040, 0016, 0002, 0000

The mid square method shown in Example 4.1 is the earliest pseudorandom number generator. It is simple and effective in execution (i.e., computationally not costly). However, based on the selected seed, it may either degenerate to zero or regenerate one of the earlier random numbers quickly. For example, when it starts with 6100, it regenerates 6100 after three random numbers. This looping behavior is inevitable for random number generators. Each loop is typically called a cycle, and the length of a cycle (i.e., the number of different random numbers generated in each cycle), is called the period of the generator. It is mathematically clear that there is a limit l for the period, which is based on the size of the random values. When the period of a random number generator is equal to the maximum possible period l, the random number generator has a full period.

A good random number generator should have the following characteristics:

- It generates numbers that do not exhibit any correlation and are distributed uniformly on $U(0, 1)$.
- It is simple and computationally effective (i.e., fast and limited memory requirement).
- It has a long and preferably full period.
- It can regenerate the same stream of random numbers when it starts with the same seed.
- It can generate different streams of random numbers when it starts with different seeds.

Most random number generators used in the simulations are linear congruential generators (LCGs), which show these characteristics. In LCG, the following recursive function generates a stream of random numbers distributed according to the uniform distribution $U(0, 1)$:

$$U_i = Z_i/m$$

$$Z_i = (a Z_{i-1} + c) mod\, m$$

where m, a, c, and Z_0 are non-negative integers, and are called:

m: the modulus
a: the multiplier
c: the increment
Z_0: the seed,

When the following conditions hold, an LCG has a full period:

- The only positive integer that divides both m and c is one.
- If there is a prime number q that divides m, then q divides also $a-1$.
- If four divides m, then four divides also $a-1$.

Because of the first condition, LCGs behave differently for $c>0$ and $c = 0$. When $c>0$, the LCG is called a mixed LCG; when $c = 0$, the LCG is called a multiplicative LCG. A mixed LCG can have a full period depending on the selection of the modulus m. For a large period, m must also be large. For making the generator computationally more effective, m and a can be selected such that

$m = 2^b$, where b is the word size for the computer
$a = 2^l + 1$, where l is a positive integer

Two mixed LCG that satisfies these conditions are as follows:

$$Z_i = (5^{15} Z_{i-1} + 1) \bmod 2^{35}$$

$$Z_i = (314,159,269\, Z_{i-1} + 453,806,245) \bmod 2^{31}$$

A multiplicative LCG is more advantageous compared with a mixed LCG because the addition of c is not needed for the multiplicative LCGs. However, a multiplicative LCG cannot have a full period. When $m = 2^b$, a multiplicative LCG can have a period of at most 2^{b-2}. To avoid this, the largest prime number less than 2^b can be selected for m. For example, when the word size is 32, then we can use $m = 2^{31} - 1$. Two examples for the multiplicative LCGs are as follows:

$$Z_i = (16,807\, Z_{i-1}) \bmod (2^{31} - 1)$$

$$Z_i = (630,360,016 Z_{i-1}) \bmod (2^{31} - 1)$$

Example 4.2

In a simulation, 20 radios are deployed randomly according to the uniform distribution in a square area of 1000 meters × 1000 meters. The coordinates of the bottom left corner of the square are (0,0). The coordinates of the other points are given by their offset from (0,0) in meters.

a. Use a multiplicative LCG to assign the coordinates for each radio.
We use the following multiplicative LCG for random number generation:

$$Z_i = (16,807\, Z_{i-1}) \bmod (2^{31} - 1)$$

Because we have two separate random variables X and Y for X and Y coordinates of a point, respectively, we will use two instances of the generator. The seeds of the instances are selected as follows:

$$Z_{X0} = 913566091$$

$$Z_{Y0} = 413682397$$

The results are given in Table 4.1. The second and third columns are the Z numbers generated by the LCG for X and Y, respectively. The fourth and fifth columns are the random numbers $U(0,1)$, where $U_{Xi} = Z_{Xi}/m$, $U_{Yi} = Z_{Yi}/m$, and $m = (2^{31} - 1)$. Finally, the sixth and seventh columns are random X and Y coordinates given by $x_i = U_{Xi} \times 1000$ and $y_i = U_{Yi} \times 1000$, respectively. Thus, $x_i = U_{Xi} \times 1000$ because x_i is distributed

Table 4.1: 20 random points.

i	Z_{Xi}	Z_{Yi}	U_{Xi}	U_{Yi}	x_i	y_i
1	1944699034	1355481040	0.90557106	0.631195	905.5711	631.195
2	2003040745	1063311904	0.93273853	0.495143	932.7385	495.1432
3	1152150843	1871743841	0.53651204	0.871599	536.512	871.5986
4	339173302	2058274431	0.15793988	0.958459	157.9399	958.4587
5	1064087576	1751775941	0.49550439	0.815734	495.5044	815.7342
6	2023561263	97440017	0.94229414	0.045374	942.2941	45.37404
7	295629702	1291826705	0.13766331	0.601554	137.6633	601.5537
8	1518726003	671759765	0.70721191	0.312813	707.2119	312.8125
9	237304179	944838076	0.11050337	0.439975	110.5034	439.9745
10	494203974	1399457414	0.23013166	0.651673	230.1317	651.6731
11	1766928069	1439855154	0.82279	0.670485	822.79	670.4848
12	1356184967	1799838882	0.63152284	0.838115	631.5228	838.1153
13	9311111	437438132	0.00433582	0.203698	4.335824	203.698
14	1873019993	1186160843	0.8721929	0.552349	872.1929	552.3492
15	2031724625	714593200	0.9460955	0.332758	946.0955	332.7584
16	58301428	1439358376	0.02714872	0.670253	27.14872	670.2535
17	619557364	2040425624	0.28850388	0.950147	288.5039	950.1472
18	1899896092	267103625	0.88470806	0.12438	884.7081	124.3798
19	619271001	969803145	0.28837053	0.4516	288.3705	451.5998
20	1381960445	80577285	0.64352548	0.037522	643.5255	37.52172

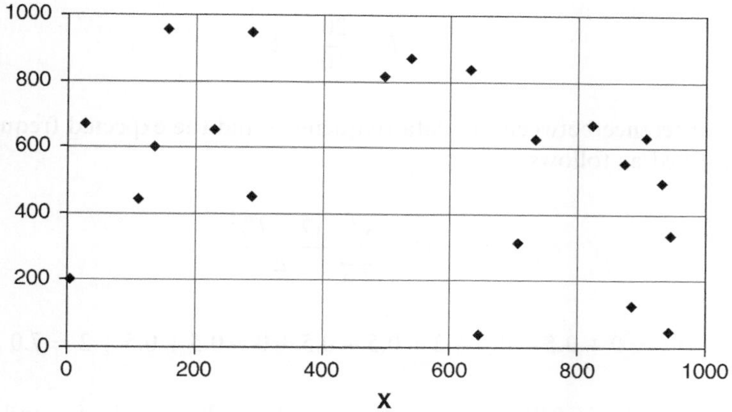

Figure 4.1: Twenty random points.

between 0 and 1000. We will elaborate on this more when explaining the realization of a random variable for the uniform distribution in the following section.

The graphical representation of the coordinates for the random points is depicted in Figure 4.1.

b. Use the χ^2 test to analyze and prove that the generated random numbers are distributed according the uniform distribution.

Let us perform this example only for the X coordinate. The solution for the Y coordinate will be similar.

Our null and alternative hypotheses are as follows:

H_0: The X coordinates for radios generated by the LCG are distributed according to the uniform distribution.

H_a: It is NOT distributed according to the uniform distribution.

We can divide the data into 10 classes (i.e., one class for every 100 meters) as follows:

Class	0–99	100–199	200–299	300–399	400–499	500–599	600–699	700–799	800–899	900–999
Frequency	2	3	3	0	1	1	2	1	3	4

We use the significance level $\alpha = 5\%$, and the degrees of freedom $v = 9$. The critical value for $\chi^2_{5\%}(9)$ is

$$\chi^2_{5\%}(9) = 16.919$$

Since we are testing for the discrete uniform distribution, the expected frequency value is

$$E = \frac{20}{10} = 2$$

The difference between the data frequencies and the expected frequencies is quantified as follows:

$$X^2 = \sum_{i=1}^{n} \frac{(O - E)^2}{E}$$

$$X^2 = 0 + 0.5 + 0.5 + 2 + 0.5 + 0.5 + 0 + 0.5 + 0.5 + 2 = 7.0$$

Because $7.0 < 16.919$, we cannot reject the null hypothesis, and we can assume that the observed data are distributed randomly according to the uniform distribution. This conclusion is significant at the 5% level.

4.1.2 Realization of Random Variables for a Simulation

In the previous section, we learned how to generate random numbers distributed between zero and one according to the uniform distribution. Of course, we need more than this for the realization of random variables. The question is how to generate random values according to a given discrete or continuous distribution and parameters. Various techniques are used to generate a function that can do this, such as inverse transform, composition, convolution, and acceptance-rejection. These techniques are out of the scope of this book. Interested users can find the details about these techniques in [LAW91]. In Table 4.2, the algorithms to generate random values according to various distributions are given.

Table 4.2: Algorithms for random variable realization.

Distribution	Parameters	Algorithm
Bernoulli	p: the probability of success	– Generate a random number u, which is $U(0,1)$. – If $u \leq p$ return "***success***", otherwise return "***fail***".
Discrete uniform	a: the index for the lowest class b: the index for the highest class Outcomes are distributed in the classes from a to b.	– Generate a random number u, which is $U(0,1)$. – Return $a + \lfloor (b-a+1)\, u \rfloor$.
Binomial	p: the probability of success t: the number of trials The number of successes in t trial.	– If Bernoulli(p) = success, $x_1 = 1$. Otherwise $x_1 = 0$. – If Bernoulli(p) = success, $x_2 = 1$. Otherwise $x_2 = 0$. – If Bernoulli(p) = success, $x_t = 1$. Otherwise $x_t = 0$. – Return $x_1 + x_2 + \cdots + x_t$
Geometric	p: the probability of success The number of trials before the first success.	– Generate a random number u, which is $U(0,1)$. – Return $\lfloor \ln u / \ln(1-p) \rfloor$.
Negative binomial	p: the probability of success s: the number of trials The number of trials before the sth success.	– x_1 = Geometric(p) – x_2 = Geometric(p) – x_s = Geometric(p) – Return $x_1 + x_2 + \cdots + x_s$.
Poisson	λ: rate	– $a = e^{-\lambda}$ – $b = 1$ – $i = 0$ – Generate a random number u, which is $U(0,1)$. – $b = b\,u$ – if $b < a$ return i

(Continued)

Table 4.2. Continued

Distribution	Parameters	Algorithm
		– $i = i + 1$ – Go to the fourth line.
Uniform	*a*: the lowest value *b*: the highest value Numbers are distributed between *a* and *b*.	– Generate a random number *u*, which is $U(0,1)$. – Return $\boldsymbol{a + (b - a)\, u}$
Exponential	μ: mean value	– Generate a random number *u*, which is $U(0,1)$. – Return $\boldsymbol{-\mu \ln u}$
Normal	μ: mean value σ: standard deviation	– Generate a random number u_1, which is $U(0,1)$. – Generate a random number u_2, which is $U(0,1)$. – $V_1 = 2U_1 - 1$ – $V_2 = 2U_2 - 1$ – $W = V_1^2 + V_2^2$ – If W > 1, go back to the first step – $Y = \sqrt{(-2\ln W)/W}$ – Return $\boldsymbol{\mu + \sigma V_1 Y}$ and $\boldsymbol{\mu + \sigma V_2 Y}$ – Note that the algorithm returns two random values at each run.

Example 4.3

For the same problem in Example 4.2, distribute the radios according to the normal distribution with mean $\mu = 200$ and standard deviation $\sigma = 50$. Use the same LCG with the same seeds. Compare the graphical depiction of the radios distributed according to the uniform and normal distributions.

Table 4.3: Twenty random points distributed according to the normal distribution.

i	Z_{Xi}	Z_{Yi}	U_{Xi}	U_{Yi}	X	Y
1	1944699034	1355481040	0.905571	0.631195		581.7132
2	2003040745	1063311904	0.932739	0.495143		496.975
3	1152150843	1871743841	0.536512	0.871599	504.5895	
4	339173302	2058274431	0.15794	0.958459	457.0032	
5	1064087576	1751775941	0.495504	0.815734	499.7484	
6	2023561263	97440017	0.942294	0.045374	524.7557	
7	295629702	1291826705	0.137663	0.601554	473.9174	531.1518
8	1518726003	671759765	0.707212	0.312813	514.916	442.5799

(*Continued*)

Table 4.3. Continued

i	Z_{Xi}	Z_{Yi}	U_{Xi}	U_{Yi}	X	Y
9	237304179	944838076	0.110503	0.439975	486.5297	472.4609
10	494203974	1399457414	0.230132	0.651673	490.6669	569.5862
11	1766928069	1439855154	0.82279	0.670485	539.3342	516.7843
12	1356184967	1799838882	0.631523	0.838115	516.027	533.2877
13	9311111	437438132	0.004336	0.203698		450.377
14	1873019993	1186160843	0.872193	0.552349		508.7671
15	2031724625	714593200	0.946096	0.332758		457.3857
16	58301428	1439358376	0.027149	0.670253		543.3817
17	619557364	2040425624	0.288504	0.950147	487.7137	
18	1899896092	267103625	0.884708	0.12438	522.3486	
19	619271001	969803145	0.288371	0.4516	452.0775	498.0174
20	1381960445	80577285	0.643525	0.037522	532.5007	481.0559
21	1573556810	1347731385	0.732744	0.627586	527.8016	513.7276
22	508192865	1811362786	0.236646	0.843482	468.542	536.9567
23	655017936	846164430	0.305016	0.394026	440.5712	444.6551
24	885275830	848864576	0.412239	0.395283	473.2513	445.3117
25	1064168394	1133061811	0.495542	0.527623	494.2049	506.0099
26	1234385742	1632359528	0.574806	0.760127	597.2426	556.5955

The results are given in Table 4.3. The second and third columns are the Z numbers generated by the LCG for x and y, respectively. The fourth and fifth columns are the random numbers $U(0,1)$, where $U_{Xi} = Z_{Xi}/m$, $U_{Yi} = Z_{Yi}/m$, and $m = (2^{31}-1)$. Finally, the sixth and seventh columns are random x and y coordinates distributed according to the normal distribution.

The graphical representation of the coordinates for the random points is depicted in Figure 4.2. Note that the minimum and maximum values for the x and y axes are different from the ones in Figure 4.1.

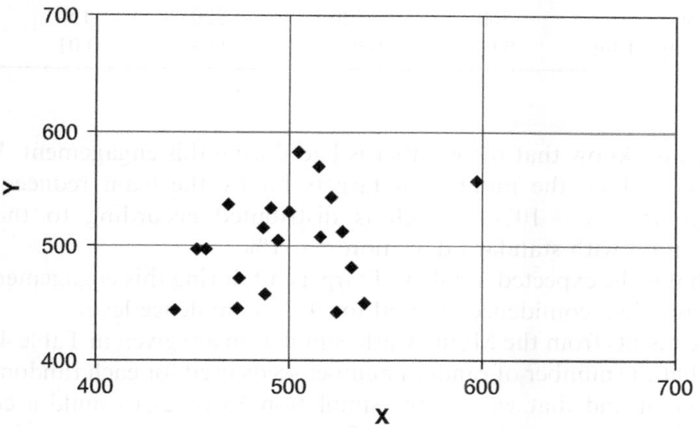

Figure 4.2: Twenty random points distributed according to the normal distribution.

4.2 STATIC SIMULATION

Static simulation is used for solving problems not analytically tractable and that do not have any relation with time. In this section, it is worth mentioning about Monte Carlo simulation, which is often defined as any simulation that involves random numbers. This definition implies that the term "Monte Carlo simulation" is synonymous with the term "stochastic simulation." A more restrictive definition of Monte Carlo simulation is that it is a simulation for solving a deterministic or stochastic problem where time does not play a role. This definition makes Monte Carlo and static simulations anonymous to each other. In this section, we restrict the definition of the Monte Carlo simulation even more. It is a type of simulation that is both static and stochastic.

The Monte Carlo simulation can be used to solve various types of static problems not analytically tractable. For example, the critical values for a goodness-of-fit test can be determined or the integral for a function not analytically tractable can be evaluated by using the Monte Carlo simulation. Example 4.4 is an application of the Monte Carlo simulation for a stochastic and static problem.

Example 4.4

A team of five troops is in engagement. Every troop has different capabilities and skills. Therefore, the probability of a hit is different for each of them as given in Table 4.4. The number of times that they are shot at is also given in Table 4.4. Note that the number of hits is distributed according to the binomial distribution.

Table 4.4: The probability of hit and number of shot values for the team.

	Troop 1	Troop 2	Troop 3	Troop 4	Troop 5
t: # of shots	300	220	230	150	200
p: probability of hit	0.01	0.005	0.008	0.01	0.006

We also know that the weather is bad during this engagement. When the weather is bad, the number of targets hit by the team reduces 10% on average (i.e., $\mu = 10\%$), which is distributed according to the normal distribution with standard deviation $\sigma = 1\%$.

What is the expected number of targets hit during this engagement? Please also provide a confidence interval for 95% confidence level.

The results from the Monte Carlo simulation are given in Table 4.5. Please note that the number of random number seeds used for each random variable is different and that we run the simulation 35 times to build a confidence interval. On average, the number of hits for this engagement is eight.

Also note that we selected the number of times as 35. As this number increases, the confidence interval gets shorter.

Table 4.5: The Monte Carlo simulation of the engagement.

	Troop 1	Troop 2	Troop 3	Troop 4	Troop 5	Weather factor	Total hits
1	1	0	3	3	3	0.1144	8.86
2	3	2	3	2	1	0.0927	9.98
3	4	0	3	0	1	0.0916	7.27
4	2	0	3	0	1	0.1046	5.37
5	2	3	4	1	3	0.0986	11.72
6	3	1	3	1	0	0.0847	7.32
7	4	3	0	0	0	0.0981	6.31
8	4	2	1	3	3	0.0929	11.79
9	1	1	5	1	0	0.1157	7.07
10	4	1	5	3	1	0.0960	12.66
11	1	2	1	0	0	0.0949	3.62
12	5	1	4	0	0	0.0955	9.04
13	4	0	6	2	0	0.1087	10.70
14	4	0	1	1	1	0.0991	6.31
15	2	1	3	2	1	0.0996	8.10
16	4	4	2	1	2	0.0872	11.87
17	5	1	2	1	1	0.1070	8.93
18	2	0	2	3	1	0.0914	7.27
19	2	0	2	0	1	0.0972	4.51
20	3	1	2	1	2	0.1014	8.09
21	4	1	4	2	0	0.1000	9.90
22	4	2	0	0	2	0.1086	7.13
23	2	1	2	0	4	0.1008	8.09
24	2	0	1	0	1	0.1054	3.58
25	3	1	1	2	0	0.1097	6.23
26	3	3	2	1	0	0.1134	7.98
27	3	0	1	2	1	0.1020	6.29
28	4	2	2	1	1	0.1006	8.99
29	1	1	1	0	1	0.0980	3.61
30	0	2	2	5	0	0.0927	8.17
31	3	1	1	0	2	0.1046	6.27
32	2	1	2	2	1	0.0856	7.32
33	3	0	1	1	0	0.0846	4.58
34	1	0	3	1	0	0.1158	4.42
35	4	2	1	2	0	0.0876	8.21

We are 95% confident that the number of hits for this engagement will be between 6.83 and 8.46.

4.3 DYNAMIC SIMULATION

In a dynamic simulation, the system under study changes its status over time. Therefore, we cannot obtain results independent from time as we did in Example 4.4 for the Monte Carlo simulation. It is possible to categorize dynamic simulation systems even more into three broad classes as discrete event, continuous, and hybrid. The main difference among these categories is related to the time advance mechanisms, which we can classify as follows:

- ***Next event time***: Time is advanced from the time of one event to the time of the next event scheduled.
- ***Fixed time increment***: In this case, time is increment with fixed intervals, such as a second, a minute, an hour, a day, and so on.

4.3.1 Discrete Event Simulation

Discrete event simulation uses the next event time advance mechanism. The events are scheduled as they are determined and put into a queue. Then, the simulation is forwarded to the first event in the queue. At that point, computations are made, and the status of the system, which is determined by state variables, is updated according to the results of these computations. This approach may cause the creation of new future events, which are scheduled and placed into the event queue according to their time.

For example, we can simulate an air-to-ground attack mission by using discrete event simulation. The events in this mission can be as follows:

- Take off
- Arrive to the target area
- Fix bomb
- Leave the target area
- Land

The events can be determined in more detail if higher fidelity is required. When it is the turn of one of these events, all the computations are carried out. For example, when it is the time for the "fix bomb" event, the following can be done:

- Calculate the impact on the target.
- Determine whether any aircraft losses have occurred.
- Recalculate the timings for the events "leave the target area" and "land." If the times change, then arrange the event queue.
- Determine whether a new event exists because of a change in the status of the simulation. If there is, schedule them.

Hence, a discrete event simulation moves from one event to the other and updates the status of the simulated system at the discrete event times.

4.3.2 Continuous Simulation

In the continuous simulation state, variables change continuously. Basically, a fixed time advance mechanism is used for continuous simulation. The fixed time increments can be very short (e.g., a nanosecond). Every time the simulation time is forwarded, all the computations are carried out as required. For example, the location and status of the aircrafts in an air-to-ground attack can be calculated at every second or minute when simulating an air-to-ground mission.

Continuous simulation can also be mixed with discrete event simulation, where time is advanced in fixed time intervals, as long as there is not a scheduled event between two fixed points in time. Let us go back to our air mission example. We can use fixed time increments where the time intervals are in 1 minute. Apart from this, we can also maintain an event queue. When the time for an event comes before the next fixed point, such as a "fix bomb" event before the next minute, time is advanced to the event point, which implies a shorter interval than the fixed interval, and computations are carried out.

4.4 PHASES IN A SIMULATION

We will follow the approach in [LAW91] for the explanation of the phases in a simulation study. Please note that these phases are not always followed. We will comment on this at the end of the section. A simulation study is typically made up of the following phases:

- ***Formulating problem and planning the study***: A good formulation and planning of the study is important for the success of the study. In this stage, if the simulation study is not for training but for an analysis task, then a careful design of the experiment is also essential. The design of experiment is explained in Chapter 6.
- ***Collecting/analyzing data and developing models***: At this stage, data about the processes and the system under study are collected and analyzed. Based on the results from these studies, models are developed.
- ***Validation of the models***: Each model should be validated carefully by the help of the experts and by using statistical data.
- ***Implementation of the models as a computer simulation***: When models are validated, they can be implemented. Various simulation systems and libraries are available. They can be used to ease this phase.
- ***Verification and validation of the simulation***: Verification ensures that the models are implemented correctly and bug free. Then, the simulation system is validated as a whole. This simulation can be run by using example data.
- ***Development and validation of the simulation database***: Almost all the previous phases are often reusable. However, each exercise and experiment typically requires its own database, although parts of databases are also generally reused. All input data should be verified carefully by using tests like goodness of fit, and their distributions and other characteristics must be validated.
- ***Running the simulation***: If the simulation is an interactive simulation that involves multiple operators, the synchronization of the operators and

accuracy in the inputs during simulation is another point that should be planned carefully. Apart from this, the determination and removal of warm-up and cool-down periods from the simulation results is another critical issue especially when the simulation is for experimentation and analysis purposes. For example, if the contention on a road examined, then it may take some time until the traffic on that road becomes typical after the simulation starts, because various convoys can be started from various points, and it may take some time to reach to the studied road. Until that point, the road may be perceived as empty, which may not be realistic. The period until traffic is as heavy as in a typical case on the road is called a warm-up period for the simulation. The same thing may happen at the end of the simulation. The part at the end of the simulation is called a cool-down period. The warm-up and cool-down periods can be removed from the simulation results by either analyzing the output data and determining the warm-up and cool-down periods or starting and ending the simulation with a realistic status.

- ***After action analysis***: The data to be collected during the simulation should be planned carefully before the simulation and collected during the simulation. If the simulation is for experimentation or analysis purposes, multiple runs may occur in which the random number seeds changed each time. At the end, output data should be categorized and analyzed by using the statistical techniques explained in Chapter 3.
- ***Publishing the final report***: A well-structured and readable report is as important as running the simulation.

Some of these phases may be skipped based on several factors. For example, when the simulation is for an exercise and the simulation system has already been validated, the phases of the simulation study can be limited to the following:

- Development and validation of the simulation database
- Running the simulation
- After action analysis
- Publishing the final report

If a federation is used for simulation, a federation development process (FEDEP) should also be run. We will explain the details about FEDEP in the next chapter. Finally, in an exercise, the process should be in line with the exercise process. We will elaborate this statement in the second part of the book.

4.5 REVIEW QUESTIONS

4.1 What is simulation?

4.2 What are the main differences among simulation systems categories?

4.3 When is static simulation used?

4.4 Generate a full cycle of random numbers by using the mid square method with the seed 2341.

4.5 For Example 4.2, generate 20 random points according to the exponential distribution.

4.6 Run the Monte Carlo simulation for 20 times. Each time, deploy 100 targets randomly according to normal distribution in a field that is 1000×1000 the same as the deployment in Example 4.2. Targets in the square with coordinates $x_{\mathrm{tl}} = 50$, $y_{\mathrm{tl}} = 50$ for top left corner and $x_{\mathrm{br}} = 100$, $y_{\mathrm{br}} = 0$ for bottom right corner are hit. Build a 95% confidence interval for the average number of targets hit each time.

4.7 Define the Monte Carlo simulation.

4.8 Compare discrete event simulation with continuous simulation.

4.9 Explain the phases in a simulation.

5

DISTRIBUTED SIMULATION

Historically, defense simulation has been confined to a single, isolated application developed solely for a single purpose. This purpose could be to train a pilot to fly a new aircraft or to simulate the trajectory of a missile as it leaves a weapon system and tracks a target through a variety of environmental conditions. The ever increasing advancement of computer-based technology has now observed many once isolated systems, which are connected together to form a complex "system of systems" world. Computer-assisted exercises (CAX) can use these same connections to help us to understand, train for, and analyze the complexities of modern warfare. It can also represent the real systems and the operational environments they perform in a simplified form that can be understood more readily. These simplified real-world representations can then be networked via participating platforms (i.e., embedded simulation on planes, ships, and tanks), in a single building (such as Maritime HQ), or even a single classroom for joint warfare training.

Baker [BAK07] has written that simulations can be divided into several types, both by their purpose and by the time-management approach that they employ. Three main purposes for simulations are training, analysis, as well as test and evaluation. Four of the main time-management approaches for simulation are real time, time stepped, event driven, and optimistic time warp. In optimistic time warp, calculations are made ahead of the current simulation time, and if events arrive in the calculations past, then the calculations are wound back to the time of the event that arrived. For most types of analysis, two time aspects are important. Any analysis scenario run on a simulator should be repeatable and should run as fast as possible. Some approaches to time management lead to

Computer Assisted Exercises and Training: A Reference Guide, by Erdal Çayirci and Dusan Marincic

minor time anomalies, such as a weapon firing after it has hit its target. If time anomalies are caused by network latency, then they will occur randomly and scenarios will not be repeatable. Nonrepeatability is a problem when a scenario is rerun with some change made. If the results of the two scenarios are different, then it is very difficult or impossible to tell whether the difference is caused by the change or by the nonrepeatability. Time is especially important in distributed simulations used for analysis because of causality and repeatability. In a simulation that runs on a single process, time is easy to deal with. Regardless of how time is dealt with, a single process can ensure that events occur in a logical sequence corresponding to real events being simulated.

For the occurrence of events, Popper [POP02] stated that the calculus of probability is a theory of certain chance-like or random sequences of events or occurrences (i.e., of repetitive events such as a series of throws with a die). These sequences are defined as "chance like" or "random" by means of two axiomatic conditions: the axiom of convergence and the axiom of randomness. If a sequence of events satisfies both of these conditions, then it is called a "collective." A collective is a sequence of events or occurrences that in principle can be continued indefinitely. For a simulation that runs in multiple processes, time is not a problem if the processes are running synchronously that is, if a mechanism exists to ensure that each process advances time in sync with every other process. For a simulation that runs in multiple processes, time is not a problem if the processes are running synchronously, that is, if a mechanism exists to ensure that each process advances time in sync with every other process. However, it can be problematic if multiple simulation processes advance time asynchronously. Under such circumstances, an event simulated by one process can occur in the past of part of the simulation running in another process. More complications can develop when multiple simulation processes are running on separate processors that are connected by a network. In this case, another problem of network latency must be dealt with.

There is a growing interest in geographically distributed training/exercising using distributed simulation, regardless of above the mentioned challenges. Recently, applications have been published in the military, space, and civil aerospace domain. Among the reported advantages are the (new) possibility to perform team training, the possibility to include real entities in the simulation, and cost reduction by saving on travel and subsistence. To exploit the advantages of fully distributed simulation exercises, three major elements are required: a standardized intercommunication mechanism and a standardized process for federation development and execution as well as exercise management.

Every time we wish to build a simulation[1] to represent a complex activity, it makes sense first to build smaller simulations to represent individual entities and

[1] Simulations are composed of five main components: simulation engines/infrastructure that execute the simulation; models that provide the real-world representations; data that provide the essential input to the models; visualization/interaction that displays the results of the simulation; and communication that provides the interfaces between simulations as well as to real-world systems.

then to make these smaller simulations interact with each other to create the desired larger simulation while spreading the computational load. It also makes sense that if we build simulations at a later date, then these simulations can interact with other existing simulations as required. It is important to recognize that the war fighter requires a capability to combine a multitude of individual simulations into larger simulations to meet many diverse requirements, which range from planning to mission rehearsal. The key enabler for this to occur is a highly adaptive simulation architecture with supporting interfaces and protocols. Two distributed simulation communication standards are currently used in the defense industry to achieve this aim. These are the Distributed Interactive Simulation (DIS) and High-Level Architecture (HLA) standards [MCF04].

A distributed simulation has multiple modules that can be run on multiple processors. All modules work together collaboratively and interactively. The processors can be colocated in the same geographic location or in remote sites. Distributed exercises are often perceived to be the same as distributed simulations, which is not true. An exercise where the training audience can be at different locations (i.e., different cities, countries or continents) because of operational, technical, or financial reasons is called a distributed exercise, and it can be supported by distributed or centralized simulations. Terminals that can access a centralized simulation can be located also on remote sites, which do not make the simulation a distributed one. When entities of the same synthetic environment are simulated in multiple software modules that can be run in multiple computers, and these entities modeled in different modules can interact with each other, it is called a distributed simulation.

In distributed simulation, multiple modules (i.e., simulation software) called as federates interact with each other through a common architecture and protocol. The distributed system of these federates is called as federation. Distributed simulations enable software reusability and interoperability, enhance scalability, and introduce many advantages, which include the following:

- Federates can be developed by independent software developers (i.e., interoperability), which makes the users more independent from the software developers, and the field can become more open for competition.
- New federates can be added to an existing federation (i.e., reusability).
- The details of each federate can stay proprietary although federates become a part of a federation.
- Each federate works on its entities in the level of detail required for that specific federate. Only the interactions among the entities in different federates are exchanged in the federation. Therefore, more complex and larger synthetic environments can be simulated (i.e., scalability).
- Different requirements can be fulfilled within the same federation. For example, multiple resolutions can be provided (e.g., one federate simulates in theater level, whereas another works on a portion of it with a higher resolution). Hence, multiple echelons can be simulated with various levels

of abstraction. Different aspects of a theater can also be modeled by different federates. For example, one federate can focus on logistics and another on the civil issues.

- The modular approach is often more cost effective and flexible. Instead of designing and developing a big monolithic simulation from scratch, multiple and simpler federates that can be integrated into a federation with already existing ones are generally cheaper and more flexible.

Because of these and many other reasons, distributed simulation has been investigated extensively for military purposes since the second half of 1980s. The standard itself is patterned very closely after the original simulator networking (SIMNET) distributed interactive simulation protocol, which was developed by Bolt, Beranek, and Newman for Defense Advanced Research Project Agency (DARPA) in the early through the late 1980s. They introduced the critical concept of dead reckoning to transmit the state of battle field entities efficiently; The authors also implemented DARPA's vision of simulations that involve inexpensive general-purpose computers and hundreds of online players, wherein the realism and training value came not from high-fidelity simulation of vehicle dynamics but by the real-time play with intelligent allies and intelligent opponents.

SIMNET was one of the early efforts in this direction, and it connected simulators in the remote sites. SIMNET was the proof of the distributed simulation concept for the military, and it was followed by two important projects of the field: distributed interactive simulation (DIS) and aggregate level simulation protocol (ALSP). ALSP was used for a U.S. Army federation called joint training confederation (JTC), which later became the joint land component constructive training capability (JLCTC) that uses a distributed simulation architecture different from ALSP, namely high-level architecture (HLA). HLA basically merges the ideas in DIS and ALSP. Contemporarily HLA and DIS are the most widely used distributed simulation standards for the military and in some civilian applications. Although HLA is the successor of not only ALSP but also DIS, and HLA compatibility is mandatory for military simulations in many countries, DIS is still around and being used in various federations. Therefore, we will dedicate a section also for DIS in this chapter.

Some commercially distributed architectures are not intended to be developed for military simulations, such as common object request broker (CORBA) by the object management group (OMG) and remote method invocation (RMI) from Sunsoft's Java Development Kit (JDK), but these technologies can be used for distributed simulation purposes. We exclude them from the content of this book. Interested readers can find a comparison of CORBA and RMI with HLA in [BUS08].

5.1 DISTRIBUTED INTERACTIVE SIMULATION

DIS is an open standard for conducting real-time, platform-level war gaming across multiple host computers, and it is used worldwide especially by military organizations and by other agencies, such as those involved in space exploration and medicine. DIS connects arms training and other simulation systems so that they can participate in the same simulation exercise, which allows team training to take place. DIS extended the concept of SIMNET to networking simulators of differing manufacture and different fidelity and function.[5] Funding and research interests for DIS standards development decreased following the proposal and promulgation of its successor, the HLA (simulation) in 1996. A North Atlantic Treaty Organization (NATO) standardization agreement (STANAG 4482, *Standardized Information Technology Protocols for Distributed Interactive Simulation (DIS)*, adopted in 1995), was created on DIS for modeling and simulation interoperability, but this was also abandoned in favor of HLA as early as 1998. The first draft HLA STANAG ran afoul of administrative procedures when it changed sponsors within NATO, which forced the process to start all over again at square one. In 2006, the HLA STANAG (4603) was finished and ratified by several NATO nations [BAK07].

DIS is an open standard developed to provide an infrastructure through which geographically distributed real-time simulations can interact by using the same simulation time. Three basic ideas/concepts can summarize the DIS architecture: protocol data units (PDUs), peer-to-peer heartbeat, and dead reckoning.

PDU: In a DIS exercise, distributed simulations exchange information by broadcasting standardized PDU. The version of DIS protocol known to us when this book is written defines 67 PDU types categorized into 11 families as follows:

- PDU types for Family 1 (***entity information/interaction***): entity state, collision, collision-elastic, and entity state update
- PDU types for Family 2 (***warfare***): fire and detonation
- PDU types for Family 3 (***logistics***): service request, resupply offer, resupply received, resupply cancel, repair complete, and repair response
- PDU types for Family 4 (***simulation management***): start/resume, stop/freeze, acknowledge, action request, action response, data query, set data, data, event report, comment, create entity, and remove entity
- PDU types for Family 5 (***distributed emission regeneration***): electromagnetic emission, designator, underwater acoustic, IFF/ATV/NAV AIDS, and supplemental emission/entity state
- PDU types for Family 6 (***radio communications***): transmitter, signal, receiver, intercom signal, and intercom control
- PDU types for Family 7 (***entity management***): aggregate state, IsGroupOf, transfer control request, and IsPartOf
- PDU types for Family 8 (***minefield***): minefield state, minefield query, minefield data, minefield and response negative acknowledgment
- PDU types for Family 9 (***synthetic environment***): environmental process, gridded data, point object state, linear object state, and areal object state

- PDU types for Family 10 (***simulation management with reliability***): create-entity-R, remove-entity-R, start/resume-R, stop/freeze-R, acknowledge-R, action request-R, action response-R, data query-R, set data-R, data-R, event report-R, comment-R, record query-R, set record-R, and record-R
- PDU types for Family 11 (***live entity***): time space position information, appearance, articulated parts, LE fire, and LE detonation

Each PDU consists of a header and a body. The header shall contain fields common to all DIS PDU, and it is 96-bit long as follows:

- Protocol version (8 bits)
- Exercise ID (8 bits)
- PDU type (8 bits)
- Protocol family (8 bits)
- Time stamp (32 bits)
- PDU length (16 bits)
- Padding field (16 bits)

The body of a PDU depends on its type. For example, the body of an entity state PDU includes the following fields:

- Entity ID (48 bits)
- Force ID (8 bits)
- Number of articulation parameters (8 bits)
- Entity type record (64 bits)
- Alternate entity type record (64 bits)
- Entity linear velocity record (96 Entity type record (64 bits)
- Entity location record (192 bits)
- Entity orientation record (96 bits)
- Entity appearance record (32 bits)
- Dead reckoning parameters record (320 bits)
- Entity marking record (96 bits)
- Entity capabilities record (32 bits)
- Articulation parameters ($n \times 128$ bits)

DIS is an open Institute of Electrical and Electronics Engineers (IEEE) standard, and the format of all PDU types can be found in IEEE 1278.1a. There is also a standard for the communications architecture requirements (CARS), which is IEEE 1278.2.

Peer-to-peer heartbeat: In DIS, there is no central node that manages the federates to avoid central server latency. Instead, every federate broadcasts

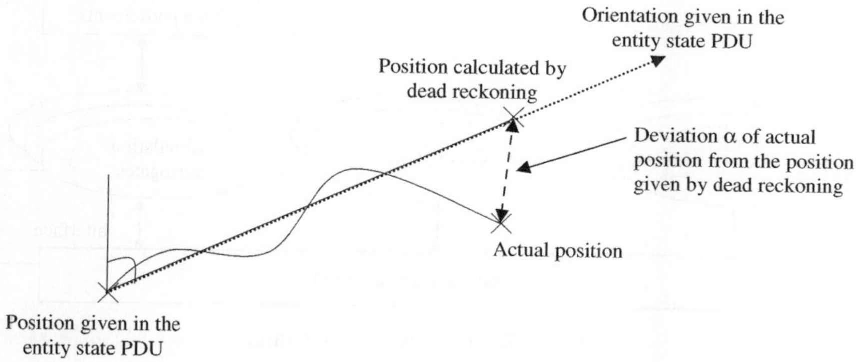

Figure 5.1: Dead reckoning in DIS.

a heartbeat, which indicates that the federate joins the federation, and it is alive. Heartbeats are broadcasted periodically, typically in every 5 seconds. In addition to the periodical ones, heartbeats can also be broadcasted when needed. Federates can be synchronized, and time can be managed by using these heartbeats.

Dead reckoning: Dead reckoning is one of the enabling concepts for DIS (Figure 5.1). Each federate receives an "entity state" PDU for the entities owned by the other federates. After receiving this PDU, federates should update the position of the entity by dead reckoning until receiving the next "entity state" PDU. Dead reckoning is typically performed based on the position, velocity, and orientation fields in the "entity state" PDU.

Federates run dead reckoning not only for the external entities but also for their own entities. The entities owned by the federate are called player entities, and the other entities are called ghost entities. For dead reckoning, there is a given threshold. Whenever the federate recognizes that the difference between the actual position of a player entity and its position determined by dead reckoning is over the threshold, the federate broadcasts an entity state PDU without waiting, for the next periodical entity state PDU. The same approach can also be applied to the other attributes of the player entities, such as strength.

The attributes calculated by dead reckoning often deviate from the actual ones. When a federate receives an entity state PDU, it can find the deviation and should correct the status of the entity accordingly. Softening algorithms achieve this without sudden jumps.

5.2 HIGH-LEVEL ARCHITECTURE

HLA is a generic and open standard for distributed simulations, and it was developed after DIS. Supposedly it replaces both ALSP and DIS. The vision of HLA is to provide an architecture that can integrate all kinds of live, virtual,

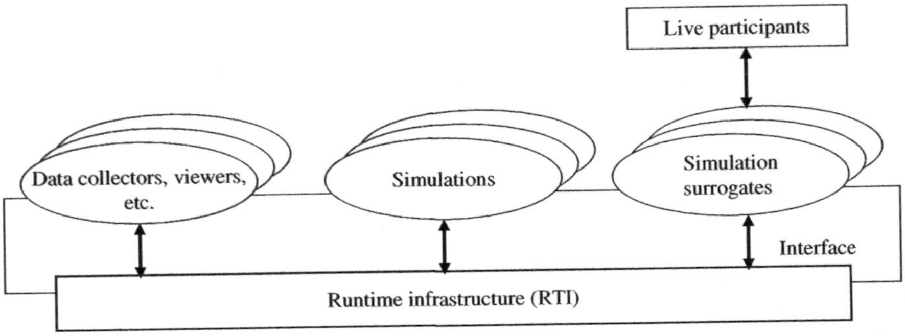

Figure 5.2: High-level architecture.

and constructive simulations and federations into a seamless distributed environment. In HLA, a central module called a run-time infrastructure (RTI) connects all federates to each other as depicted in Figure 5.2.

HLA provides specifications and standards that explain how federates can interact with each other through an RTI, and it consists of three components: object model template (OMT), HLA interface specification, and HLA rules. Apart from these three components, the HLA standard provides a standardized and recommended process for developing interoperable HLA-based federations. This process is called a federation development and execution process (FEDEP). In the following sections, we explain these three components and FEDEP in detail.

In the beginning, the development of the HLA standards was sponsored by the U.S. Defense Modeling and Simulation Office (DMSO). The final version of the HLA standards developed under the sponsorship of DMSO is DoD HLA 1.3 published in 1998. Although there are many implementations of DoD HLA 1.3, it is not internationally recognized because HLA is defined also under IEEE Standard 1516, which was released as an international standard in 2000; see the correlated documents as follows:

- IEEE 1516-2000, Framework and rules
- IEEE 1516-1-2000, Federate interface specification
- IEEE 1516-2-2000, OMT
- IEEE 1516-3-2003, FEDEP
- IEEE 1516-4-2007, Verification, validation, and accreditation (VV&A)

Later, the U.S. Department of Defense (DoD) also released the interpretations of IEEE 1516 series of HLA standards in 2003, and the Simulation Interoperability Standards Organization (SISO) developed the dynamic link compatible application protocol interface (DLC API) as a complementary specification for both HLA 1516 and DoD HLA 1.3 API specifications.

Without DLC API, the recompilation of federates is required for each different RTI implementation.

When this book was written, HLA 1516 was under revision, and the next HLA standard called HLA-Evolved was being advertised. HLA-Evolved (or HLA 1516–2009) includes DoD interpretations of IEEE 1516 and the DLC API into HLA 1516, and it has other enhancements such as follows:

- Modular Federation object models (FOMs) that will be explained in the following section
- Web services [Web Services Desinition Language (WSDL)] Application Programming Interface (API)
- Fault tolerance services that is fault (e.g., crash, network outage, etc.) notification
- Smart update rate, that is, federates can indicate how frequently they need data
- Support for data logging
- Extended support for transfer services [e.g., quality of service, Internet Protocol version 6 (IPv6), etc.]
- Extended Extensible Markup Language (XML) support for FOM Simulation object model (SOM)

The main concepts for the components specified in HLA standards are explained generically below.

5.2.1 HLA Interface Specification

The HLA interface specification defines how RTI services can be accessed by federates, and it is provided as an API in several forms, including CORBA, ADA, C++ , and Java. In the standard, there are service specifications for more than 125 RTI services defined by the following fields:

- Name and description
- Supplied arguments
- Returned arguments
- Preconditions
- Postconditions
- Exceptions
- Related services

An RTI is basically a unique implementation of HLA interface specifications, and it must support at least one programming language as defined by the RTI APIs. An RTI is made up of a central RTI component (CRC) and a local RTI

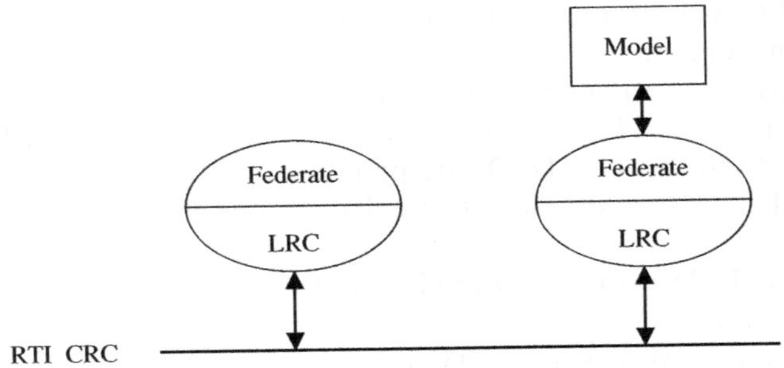

Figure 5.3: An HLA federation.

component (LRC). By using HLA API, federates communicate with the CRC through their LRC as depicted in Figure 5.3. The LRC converts RTI services accessed by a federate into messages to the other federates. The LRC also converts messages from the other federates into RTI callbacks.

A federate is compiled with its LRC, and the LRC becomes a part of the executing federate process. The LRC maintains state on the federate's behalf [i.e., saved/restored (LRC checkpoints), for known objects (reflected or owned), owned attributes, federate time, federate publication/subscription list, and undelivered message queues].

HLA services implemented in an RTI are divided into the many service groups (see Figure 5.4).

- ***Federation management***: Federates use federation management services for creation, dynamic control, modification, and deletion of a federation execution. These services include *join, resign, create, delete checkpoint, restart, pause*, and *resume*. A software system can participate in the same federation execution as multiple federates. Moreover, a software system may

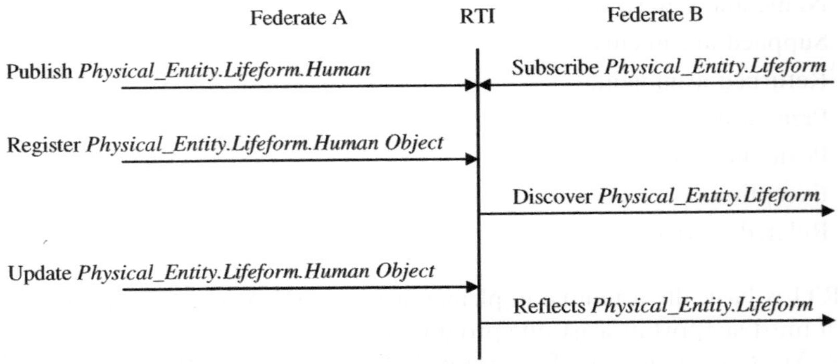

Figure 5.4: Declaration and object management.

participate in multiple federation executions. Although federation management services do not support these, they do not preclude them either.

- ***Declaration management***: Federates declare their intention to *publish* or *subscribe* to object attributes and interaction by using declarations services. *Publish* services should be invoked before registering an instance of an object. Similarly, a federate is required to subscribe to receive the updates about the object as shown in Figure 5.4.
- ***Object management***: Objects are registered on the RTI by the register object instance service. Then, objects can be modified or deleted by using object management services. The federate that registers an object owns all the attributes that are published and a special attribute called *priviledgeToDelete*. Only a federate that owns the *priviledgeToDelete* attribute can delete an object.
- ***Ownership management***: Federates can transfer the ownership of instance attributes, including *priviledgeToDelete* by using ownership management services. Only one federate can own an attribute of an object instance at a time, and only that federate can assign a value for that attribute.
- ***Time management***: Logical time is the time as represented by the simulation. No necessary connection exists between logical time and real time, which is also called wall clock time. Federates coordinate the advance of logical time and maintain its relation to the real time by using time-management services. Without these services, time synchronization should be provided externally. Without time management, it may be difficult to maintain consistency during inter-federate interactions. Four approaches to time management are as follows:
    - ***No explicit time management***, where RTI time management services are not used.
    - ***Conservative synchronization***, where no federate advances logical time until guaranteed not to receive events in the past.
    - ***Optimistic synchronization***, where federates can compute into the future, but may receive messages in the past.
    - ***Activity scan***, where federates progress through episodes in which they exchange messages at the same logical time, until they agree to advance together.

In conservative synchronization, a parameter called lookahead is needed. Lookahead is the property of a federate, not a federation, and it affects the following:

- How quickly can a model respond to an external event?
- How quickly can a model affect another?
- Does the model proceed in uniform time steps?

Federates can be either time regulating or time constraint. Time-regulating federates specify a positive lookahead. RTI does not allow the federate to send events with times less that its current logical time plus lookahead. Time-regulating federates send their messages with a time stamp. Time-constraint federates may receive messages with time stamps in the future according to their logical time and cannot prevent the federation from advancing the time.

RTI maintains two queues of events for each federate: receive order and time-stamp order. If a message is coming from a time-regulating federate and is time stamped, it is inserted into the time-stamp order queue of the federate according to time stamp. Otherwise, it is added to the end of the receive order queue from which the events are delivered to the federate without respect to logical time.

- ***Data distribution management***: Federates can reduce the transmission and receipt of irrelevant data by using data distribution management (DDM) services. DDM can help to refine the data requirements at the instance attribute level.
- ***Support services***: These services include other ancillary services required for the execution of a federation.

5.2.2 Object Model Template (OMT)

The OMT specifies what information is exchanged among federates and defines the following two types of documents to record these specifications:

- FOM is for the overall federation and describes the shared objects as well as and their attributes, interactions, and parameters for the whole federation.
- SOM has the same format as FOM, but it defines only the shared objects, attributes, interactions, and parameters used by a single federate. The SOM contains all federate information visible to other federates in a federation and all information from the other federates that may be reflected in the federate. SOM is typically a subset of FOM.

An object model (i.e., a FOM or a SOM) contains several tables, which include identification, object class, attribute, interaction, parameter, enumerated and complex data types, routing space/DDM, and lexicons. Among these tables the identification table provides a general description of the object model and consists of the fields shown in Example 5.1.

Example 5.1

Identification table

Category	Information		
Name	NATO Training Federation (NTF)		
Version	1.1		
Date	August 3, 2008		
Purpose	To create a federation where many nations join to a combined exercise by using their models.		
Application Domain	Training		
Sponsor	JPC		
POC (Title, First, Last)	Chief CAX Sup Br	Erdal	Cayirci
POC Organization	JPC		
POC Telephone	(212)2845073		
POC Email	ecayirci@jpc.org		

Object class and attribute tables are among the important tables in an object model. Objects are the simulated entities and units in a military simulation. The status of an object is represented by the values assigned to its attributes. In both DIS and HLA, an object-oriented approach is used, where an object class hierarchy provides all the attributes related to an object. An object class structure and attribute table is created by using object model development tool (OMDT) and is shown in Example 5.2. The object class shown in Example 5.2 is a subset of the NATO Training Federation (NTF version 1.0) or Joint Multi Resolution Model (JMRM version 1.0) object class model.

Example 5.2

Object class structure table

Class 1	Class 2	Class 3	Class 4
Game_Object (N)	Surface (N)	Unit (N)	Ground (PS)
			Ship (PS)
			Squadron (PS)
		Target (N)	Bridge (PS)
			Facility (PS)
	Air (N)	Air_Mission (PS)	
		Missile (N)	TBM (PS)
			Cruise_Missile (PS)
Physical_Entity (N)	Lifeform (N)	Human (PS)	

(Continued)

Continued

Class 1	Class 2	Class 3	Class 4
	Platform (N)	Aircraft (PS) Ground_Vehicle (PS) Surface_Vessel (PS) Submersible_Vessel (PS)	
Manager (N)	Federation (N) Federate (N)		

Attribute table

Object	Attribute	Datatype	Cardinality
Game_Object	Name	String	1
	Side	Side_Enum	1
	Location	Location_2D_Struct	1
Surface	Prototype_Name	String	1
	Strength	unsigned short	1
	Detect_Time	Time_Struct	1
Unit	Size	Unit_Size_Enum	1
	Mission	Unit_Mission_Enum	1
	Posture	Unit_Posture_Enum	1
	Speed	Float	1
	Course	Float	1
	Orientation	Orientation_Enum	1
	TOE_Equipment	Category_Value_Struct	1 +
Ground	Artillery_Range	Float	1
	Branch	Branch_Enum	1
	Lifting_Formation	Object_Reference	1
Ship	Type	Ship_Type_Enum	1
	Depth	unsigned long	1
	Catapult_Status	Catapult_Status_Enum	1

Note that the fields in an attribute table are more than four. The other fields in this table are units, resolution, accuracy, accuracy condition, update type, update condition, transferable/acceptable, updateable/reflectable, and routing space.

As we explained, DIS and HLA follows an object-oriented approach where a class inherits all the attributes from a class higher in its hierarchy. For example, although Attribute "*Location*" is not an attribute in the attribute list of Class "*Unit,*" Class "*Unit*" has also Attribute "*Location*" because "*Location*" is an attribute of Class "*Game_Object*," and Class "*Unit*" is a subclass of Class "*Game_Object*."

DoD HLA 1.3 provides a list of data types, which includes string, boolean, octet, short, unsigned short, long, unsigned long, long long, unsigned long long, float, double, any, and sequence. HLA 1516 adds support for a full and unambiguous specification of data types down to the bit level. Apart from these standard data types, complex and enumerated data types can also be assigned to attributes. Complex data types are composite structures where each component of the type can be defined like another attribute. Enumerated data types provide one-to-one mapping from a logical name to an integer. For an object class, the same attribute may need to be repeated multiple times. That is defined by using the cardinality field. If the attribute is singleton, then 1 is assigned to this field. If it can be none, singleton, or multiple, 0+ is assigned. If it is at least one and may be more, 1+ is assigned. Finally, a fixed value, such as 8 can also be assigned to the cardinality field.

An FOM contains information about object classes both related to scenario and to metadata. However, note that any information about the instances of these classes is not included in an FOM. An FOM describes the object data that can be shared in a federation, but it does not provide actual data instances or information about how and when they will be instantiated. That makes an FOM reusable also for the other federations.

Other object classes, exist, called manager, federation, and federate in Example 5.2. These classes are for a mechanism called management object model (MOM) by which the federation can be managed. The MOM is a set of object classes and interactions that can be used by any federate to poke, probe, and query aspects of an executing HLA federation. The rules for MOM objects and interactions are exactly the same as the other FOM object classes and interactions. By default, MOM objects are owned by the LRC (RTI) and are used for various interactions, including the following:

- Query/response
- Parameter/configuration changes
- RTI/federate diagnostic information
- Service invocation

Federations can extend the MOM to suit their needs, but the RTI will neither respond to nor understand the extensions.

An interaction is a change in the state of an object in sending a federate, which may cause also a change in the state of an object in receiving the federate. A ***parameter*** is the information associated with an interaction. An interaction is an instantaneous event shared between federates. As shown in Example 5.3, interactions structured into a hierarchy are similar to the object class hierarchy.

Example 5.3

Interaction table

Interaction 1	Interaction 2	Interaction 3	Interaction 4
Order (N)	Air_Order (N)	Offense_Order (N)	Wild_Weasel (IR)
			Air_Ground_Attack (IR)
		Defense_Order (N)	Combat_Air_Patrol (IR)
			Escort (IR)
	Ground_Order (N)	Defend (IR)	
		Attack (IR)	
		Fire_Artilery (IR)	
Manager (IR)	Federate (IR)	Request (IR)	RequestPublications (IR)
			RequestSubscriptions (IR)
		Report (IR)	ReportObjectPublication (IR)
			ReportObjectSubscription (IR)
			ReportObjectsOwned (IR)

Parameter table

Object	Attribute	Datatype	Cardinality
Order	Client_Name	String	1
	Sending_Federate	unsigned long	1
	Reference	String	1
Air_Order	Squadron_Name	String	1
	Nbr_Of_Aircraft	unsigned short	1
	Ingres_Route	Route_Point_Struct	1 +
	Egres_Route	Route_Point_Struct	1 +
Report Objects Owned	ObjectCounts	ObjectPairs	unbounded

RTI does not use all the data in FOM at runtime. A subset called federation document data (FDD) is used by RTI. The minimum FDD includes object classes, attributes, interaction classes, parameters, and some other tables. Although an FDD can have all the tables in a FOM, the RTIs are required to run with the FDD defined as the minimum in the standard.

HLA-evolved introduces many enhancements also to object models. One of these is called the modular FOM approach, which increase the flexibility during

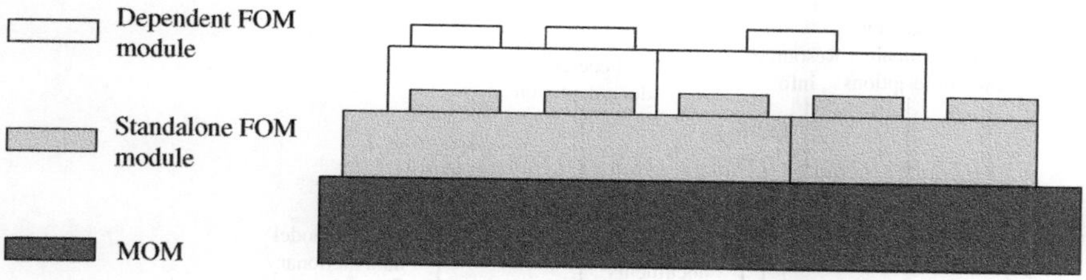

Figure 5.5: Modular FOM building blocks.

both runtime and development process. This concept allows merging multiple FOMs or extending existing FOMs without modifying them, which means modules for an FOM can be developed by several communities in different domains, or they can be can be extended by the others. With this approach, FOMs can also be decomposed into FOM modules. In the modular FOM approach, three types of modules are as follows: MOM module, standalone FOM module and dependent FOM module. A standalone FOM module can be used on its own, without any other FOM module. However, a dependent FOM module refers to a definition in another FOM module. Typically, one MOM, one or more standalone FOM, and several dependent FOM modules can create a modular FOM as depicted in Figure 5.5 [MOL07].

Please note that we exclude details like data interchange formats (DIFS) from the content of this book because we believe that those topics are implementation-level details. Interested users can find all the details in the related standards.

5.2.3 FEDEP

IEEE 1516.3 recommends FEDEP, which describes an approach for developing simulation federations. FEDEP also serves as the foundation for the verification, validation, and authentication (VV&A) overlay. Before IEEE 1516.3, FEDEP was described also for DoD HLA 1.3 sponsored by the U.S. DMSO. IEEE 1516.3 introduces a wide variety of enhancements compared with this earlier FEDEP. Many of these enhancements were drawn from the synthetic environment development and exploitation process (SEDEP) which was developed in support of EU project Euclid RTP 11.13. Figure 5.6 depicts an abstract view of the major steps within IEEE 1516.3 FEDEP.

Step 1. Define federation objectives: A set of requirements that are to be addressed through the development and execution of an HLA federation is defined, and they are transformed into a more detailed list of federation objectives.

Step 2. Perform conceptual analysis: An appropriate representation of the real-world domain that applies to the federation problem space and the federation scenario is developed. The federation objectives

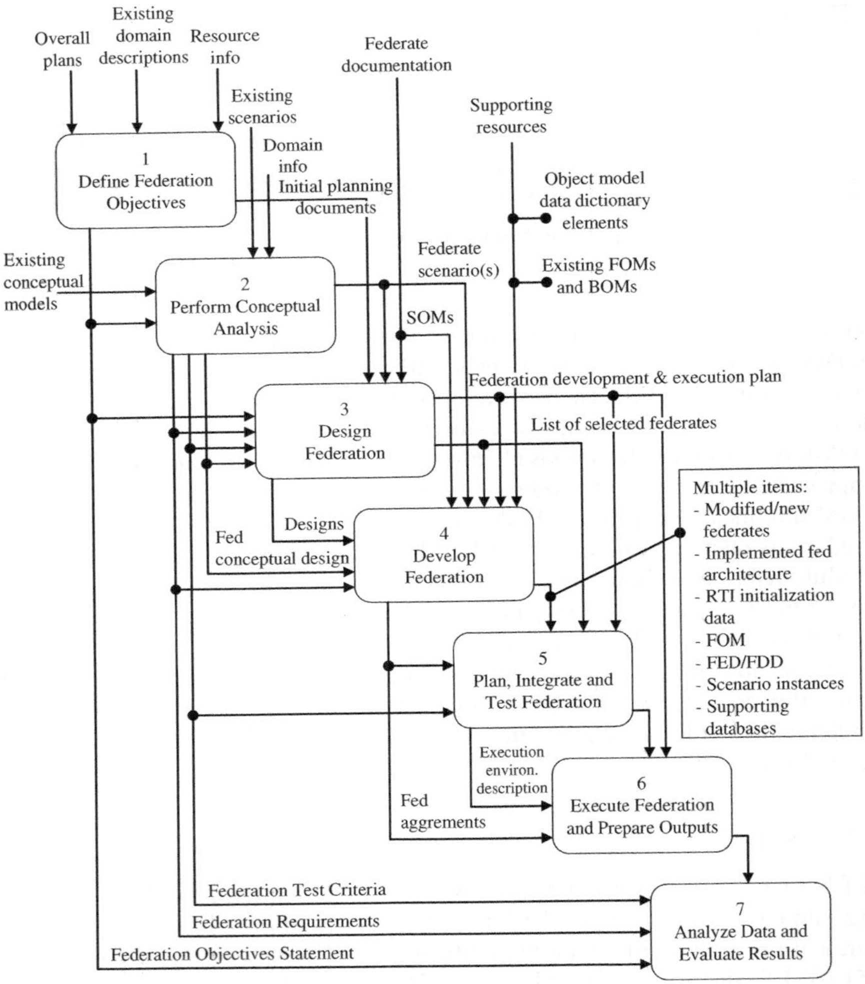

Figure 5.6: HLA 1516.3 FEDEP.

are also transformed into a set of highly specific federation requirements that will be used in federation design, development, testing, and evaluation.

Step 3. Design federation: The federation that will be implemented in Step 4 is designed. This step involves identifying existing federates that are suitable for reuse, creating new federates and federate components (if required), allocating the required functionality to federates, and developing a detailed plan for federation development and implementation.

Step 4. Develop federation: FOM is developed. The federates are modified as required, and the federation is prepared for integration and testing.

Step 5. Plan, integrate, and test federation: The federation execution is planned. All required interconnectivity among federates is established, and the federation is tested prior to execution.

Step 6. Execute federation and prepare outputs: The federation is executed, and the output data from the federation execution are preprocessed.

Step 7. Analyze data and evaluate results: The data acquired during the federation execution are analyzed, and the results are released to the users and sponsors. These results are used for evaluating how well the federations objectives are met and whether more further work is required.

5.2.4 HLA Rules

The HLA rules regulate the behavior of federations and the federates that join a federation. They are listed below as specified in DoD HLA 1.3.

Federation Rules:

- Federations shall have an HLA FOM, documented in accordance with the HLA OMT.
- In a federation, all representation of objects in the FOM shall be in the federates, not in the RTI.
- During a federation execution, all exchange of FOM data among federates shall occur via the RTI.
- During a federation execution, federates shall interact with the RTI in accordance with the HLA interface specification.
- During a federation execution, an attribute of an instance of an object shall be owned by only one federate at any given time.

Federate Rules:

- Federates shall have an HLA SOM, documented in accordance with the HLA OMT.
- Federates shall be able to update and/or reflect any attributes of objects in their SOM and send and/or receive SOM object interactions externally, as specified in their SOM.
- Federates shall be able to transfer and/or accept ownership of an attribute dynamically during a federation execution, as specified in their SOM.
- Federates shall be able to vary the conditions under which they provides updates of attributes of objects, as specified in their SOM.
- Federates shall be able to manage local time in a way that will allow them to coordinate data exchange with other members of a federation.

5.3 BASE OBJECT MODEL (BOM)

Despite the enablers (i.e., HLA, FEDEP, and APIs), one of the things that still encumbers the M&S community at large is that the task in building and putting together simulation and simulation environments, which must conform to common agreed on message interfaces, remains an arduous task. That is, it takes a long time, a great deal of effort, and a lot of collaboration. What is required is a composability infrastructure that encourages the development and reuse, across the entire community, of components that are matched to the needs of the desired simulation or simulation space. The BOM is an SISO standard (SISO-STD-003-2006), which was introduced as a key enabler for supporting composability. This section about BOM is adopted from www.boms.info, to introduce the BOM concept briefly.

BOMs are specifically identified in the IEEE 1516.3 HLA FEDEP as a potential facilitator for providing reusable model components used for the rapid construction and modification of federates and federations. The open standardization of BOM representations is considered essential for encouraging their development, distribution, and use.

The BOM concept is based on the assumption that piece-parts of simulations and federations can be extracted and reused as modeling building blocks or components. The interplay within a simulation or federation can be captured and characterized in the form of reusable patterns. These patterns of simulation interplay are sequences of events between simulation elements. The implementation of the pattern using HLA object model constructs is also captured in the BOM.

Two BOM-related documents have been standardized by the SISO via its BOM Product Development Group (PDG). These documents are as follows:

- ***The BOM Template Specification*** defines the semantics and syntax needed to represent a BOM. The BOM contains the essential elements needed to represent a reusable pattern of interplay, which is characterized by messages and/or triggers related to one or more conceptual object classes mapped to HLA object classes and HLA interaction classes.
- ***The Guide for BOM Use and Implementation*** provides discussion on BOM development and the application and use of BOMs for the assembly of simulations and simulation spaces.

Interested readers can refer to the website www.boms.info, from which SISO-STD-003-2006 can be downloaded.

5.4 REVIEW QUESTIONS

5.1 What are the characteristics of distributed simulation?

5.2 What is dead reckoning?

5.3 Explain the vision of HLA.

5.4 List the HLA interface specifications.

5.5 Describe the HLA service groups.

5.6 In HLA, who is regulating the behavior of federations and the federates and how?

5.7 What are the differences between HLA and DIS?

5.8 What are the advantages and disadvantages of having a central server like as RTI in HLA for the federation?

5.9 Create an FOM for a federation that will have federates that simulate various aspects in a university.

5.10 Is DIS or HLA more advantageous for live and virtual simulation federations? What are the advantages and disadvantages of DIS and HLA for live and virtual simulation federations?

6

EXPERIMENTATION AND ANALYSIS

In the scientific method, an experiment (Latin: exp*eriri* meaning "to try") is a set of observations performed in the context of solving a particular problem or question, to retain or falsify a hypothesis or research concerning phenomena. The Latin word *experimentum* means a trial or a test. In modern English, the verb *to experiment* is commonly defined as "any action or process undertaken to discover something not yet known or to demonstrate something known."[1] The experiment is a cornerstone in the empirical approach to acquiring deeper knowledge about the physical world.

After the end of the Cold War, the global security environment had been changed. As was explained previously, the whole system of security threats as well as the capability for their recognition and management became more complex, unpredictable, and interdependent. Like other security elements, military forces have been faced with the need to change their modus operandi to a more flexible, affordable, and manageable form. The transformation that took place in almost all armed forces in the world is still an ongoing process.

Experimentation plays a vital role in transformation. Experimentation contributes to and thus advances a body of knowledge that, when applied, allows us to develop a new capabilities. Although some think of experimentation as simply conducting individual experiments, experimentation is a *process*

[1] Webster's New World Dictionary, 1995.

Computer Assisted Exercises and Training: A Reference Guide, by Erdal Çayirci and Dusan Marincic

that combines and structures experimental results much in the way that individual bricks are fashioned into a structure for a purpose, and it steers future experimentation activities.

Descartes[2] believed that you knew what was true when all reasons for doubt were removed. Bacon[3] suggested an analytical approach. In England, a Royal Society[4] was formed to document and certify observations and methods. This was the beginning of the widespread acceptance and application of an empirically based approach to advancing understanding. Systematic observation, experimentation, and analysis now form the core of satisfying our collective curiosity and answering specific questions about "why?" and "how?" Over time, what is now known as the scientific method was developed. This accumulation of lessons learned and theory provides guidance on the design of experiments, the collection of data, its analysis, and the nature of the conclusions that can be drawn. Because any real-world observations that we make constitute a very small sample of reality under some limited conditions, statistical theory, methods, and tools play a major role in the process of knowledge acquisition [ALB05].

6.1 DESIGN OF EXPERIMENT

Alberts and Hayes [ALB02] have written that any experiment has three major phases: preexperiment, conduct of the experiment, and postexperiment. The outputs of the preexperiment phase provide "what we know" and "what we think" as expressed in the experiment model and experiment propositions in hypotheses, and "what we are going to do" as expressed in the detailed experiment plan. The output of the conduct phase is simply the empirical data generated by the experiment as well as other observations and lessons recorded. The output of the postexperiment phase is a revised model that captures and incorporates what is learned, empirical data that others can use, the documentation of supporting experimentation activities, and other findings and conclusions such as lessons learned.

The design of experiments includes the design of all information-gathering exercises where variation is present, whether under the full control of the experimenter or not. Often the experimenter is interested in the effect of some process or intervention on some objects, which may be people. The design of experiments is thus a discipline that has very broad application across all the natural and social sciences.

The general process of experimentation (Figure 6.1) begins with the order for desirable experimentation, which is created by a person placing an order or a sponsor. The order for experimentation consists of the formulation of the experimental problem, objectives, subject of experiment, parameters and

[2] Routledge Encyclopedia of Philosophy, 2004.
[3] The Internet Encyclopedia of Philosophy, 2004.
[4] The Royal Society of London, 2004.

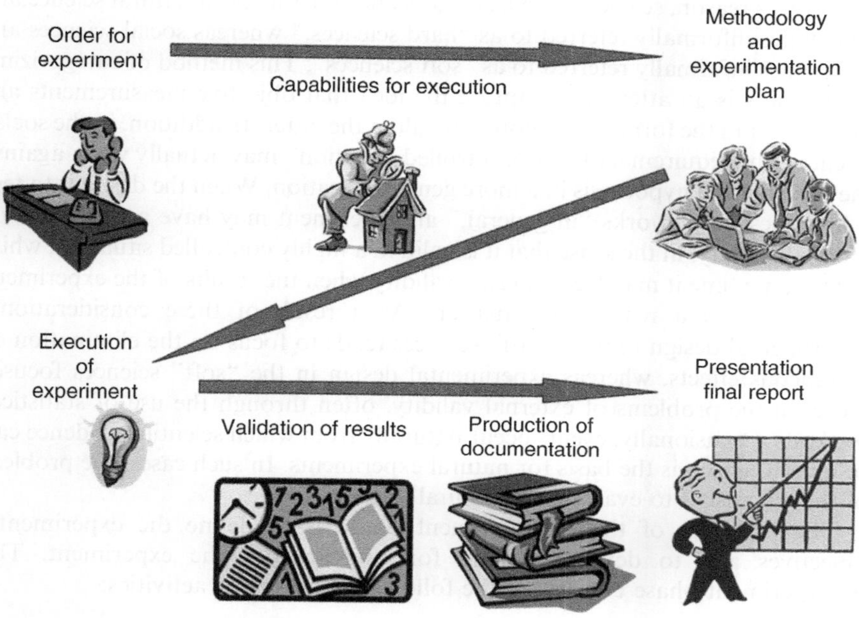

Figure 6.1: Process of experimentation.

limitations, assessment of expenditure, proposal for experimenters, and level of classification. Experimentation in military science can cover research on the tactical, operational, and strategic level of interest. After an analysis of experiment capabilities, the designated experimenter team will request additional experts and assess initial expenditure. The next step is the creation of scientific instruments, description of methodological and theoretical framework (subject, hypothesis, and aim), and establishment of experiment plan. The execution of the experiment in accordance with the detailed plan will lead to the validation of results and production of experiment documentation. The most important part is the proper presentation of findings and the reported outcomes of the experiment to the person who placed an order and to the public.

The design of experiment attempts to balance the requirements and limitations of a scientific field in which one works, so that the experiment can provide the best conclusions about the hypothesis being tested. In some sciences, such as physics and chemistry, it is relatively easy to meet the requirements that all measurements be made objectively, and that all conditions can be kept controlled across experimental trials. However, in other cases such as biology, and medicine, it is often difficult to ensure that the conditions of an experiment are performed consistently; in the social sciences, it may even be difficult to determine a method for measuring the outcomes of an experiment in an objective manner.

For this reason, sciences such as physics and other fields of natural science are sometimes informally referred to as "hard sciences," whereas social sciences are sometimes informally referred to as "soft sciences"; This method of categorizing the sciences is an attempt to capture the idea that objective measurements are often easier in the former and more difficult in the latter. In addition, in the social sciences, the requirement for a "controlled situation" may actually work against the utility of the hypothesis in a more general situation. When the desire is to test a hypothesis that works "in general," an experiment may have a great deal of internal validity, in the sense that it is valid in a highly controlled situation, while at the same time it may lack external validity when the results of the experiment are applied to a real-world situation. As a result of these considerations, experimental design in the "hard" sciences tends to focus on the elimination of extraneous effects, whereas experimental design in the "soft" sciences focuses more on the problems of external validity, often through the use of statistical methods. Occasionally, events occur naturally from which scientific evidence can be drawn, which is the basis for natural experiments. In such cases, the problem of the scientist is to evaluate the natural "design".

The objective of the preexperiment phase is to define the experiment's objectives and to develop a plan for carrying out the experiment. The preexperiment phase consists of the following four major activities:

- Formulating the experiment
- Establishing the experimentation team
- Generating the initial plan for the experiment
- Drafting the detailed plan for the experiment

Effective formulation is fundamental to the success of all experiments but particularly in transformational experiments because the issues are complex and inherently involve the many dimensions that form a mission capability package. Proper attention to formulation will provide a solid foundation on which the experiment can be built. A review of the existing body of knowledge and previous experiments will provide the team with a good idea of what is known and what conjectures have some apparent merit. The first task in formulation is to properly understand the issues that the experiment will address and the context in which the issues will be addressed. This task involves the explicit definition of several items, which include the following:

- Propositions, hypotheses, and/or relationships to be addressed
- Assumptions that will be made
- The identity of the dependent variable(s)
- The identity of the independent variables
- Which of the independent variables will be controlled
- Constraints on the value of the variables for the purpose of the experiments [ALB02]

Specific articulation of the problem is really what the experiment is all about. The assumptions made are another important part of the formulation of an experiment. They determine the scope of the experiment and clearly identify areas that are not being investigated. Making explicit assumptions is very important in helping one and all to understand and evaluate the empirical data that will result from the experimentation and its interpretation. It is always good practice to articulate clearly all the key assumptions made. The independent variables are the inputs. Taken together, they frame the experiment space. They focus us on the relationships of interest. The dependent variables are the outputs or products of an individual team or organization. They represent the characteristics and behaviors that are important to the success of military operations.

The heart of any experiment is in what we attempt to control. Experimental control can be exercised in many ways. Although the selection of variables to be controlled is part of the formulation of the experiment, how they are to be controlled is determined in detail when developing the experiment plan. The formulation task is completed with the specification of constraints that will be imposed on the variables. Some of these may be a reflection of the assumptions; others are a reflection of the scope of the experiment, or as a result of practical considerations. There are two major outputs of formulation. First is the construction of an experimentation model that contains the key variables and the relationships (some known, some hypothesized, and some the subject of discovery) between them. Second is the specification of the relationships of primary interest, the related assumptions, and constraints [ALB02].

6.2 EXECUTION OF EXPERIMENTS

The management plan for the experiment, which was developed at the beginning of the preexperimentation phase, provides the overall guidance for the experiment and specifies how the experiment will be planned and executed. In addition, this plan relates the experiment to the various external constituencies by articulating the various sponsor issues and stakeholder interests. Finally, the management plan provides the basis for securing support for the experiment from various participating military units, organizations, and other resources [ALB02].

Besides the management plan, Alberts and Heyes [ALB02] emphasized that it is necessary to develop the experiment plan as a living coordination document that communicates a consistent, comprehensive, and cohesive understanding of all the planning details to every participant, supporter, and consumer of the experiment. Developed in accordance with the management plan, the experiment plan serves as the principal control vehicle for refining and documenting agreed details of the experiment. The main elements of the experiment plan are as follows:

- Experimentation hypotheses and research questions
- Experimentation treatments, baseline conditions, and controls

- Experimentation subjects (including selection and training requirements)
- Experimentation scenarios
- Definition of measures and data collection plan
- Facility and other resource/asset requirements
- Experimentation schedule (including pretest and rehearsal events)
- Contingency options

Models and simulations have a role, albeit a different one, in support of all three uses of experiments: discovery, hypothesis testing, and demonstration. As a well-crafted experimentation campaign matures, the roles and characteristics of the models and simulations in use coevolve with changes in the operational concepts and technologies.

Model-based experimentation (Figure 6.2) has its most obvious utility in support of discovery and hypothesis testing experiments. Some specific reasons for choosing a model-based experiment include the following:

- Efficiently (in terms of both cost and time) exploring a set of scenarios, operational concepts, and technologies for combinations of circumstances that seem to be fruitful opportunities for further exploration using human subjects and hardware (discovery).

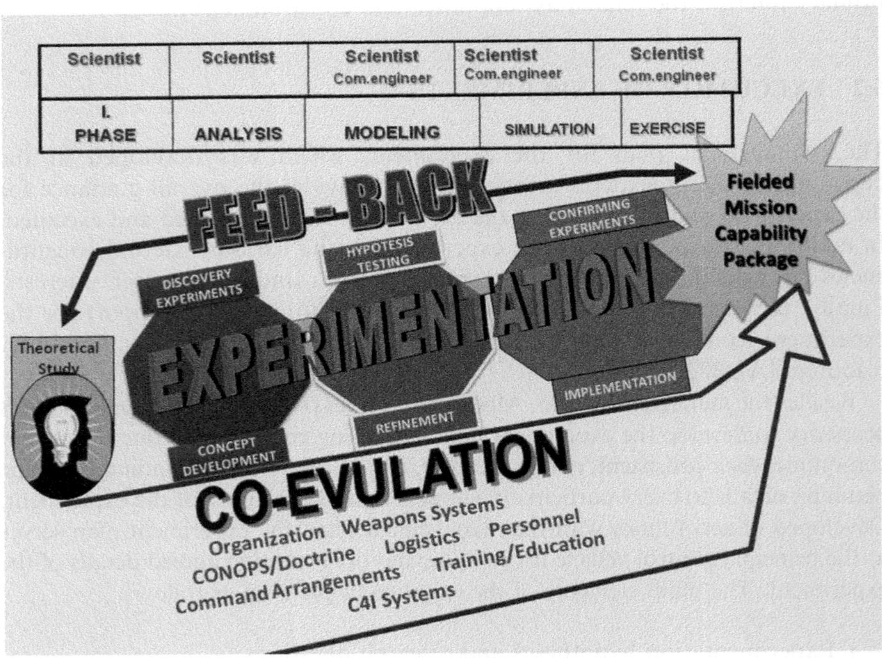

Figure 6.2: Model-based experimentation.

- Supplementing other experiments with simulation experiments that evaluate the concepts and technologies under a broader set of conditions. These data can help to inform more completely the additional experimentation, acquisition, doctrine, and force structure decisions (discovery and hypothesis testing).
- Evaluating the results of a tested concept in a different context.
- Exploring the potential utility of technologies that may not yet exist. An example is the use of small agile robots as an integral part of urban combat operations (discovery).
- Efficiently evaluating the suitability of proposed metrics (dependent variables) that are being considered for application in an experiment or experimentation campaign (discovery and hypothesis testing) [ALB02].

Model-based experimentation allows preparation of the fielded mission capability package for the operations. At same time, an experiment coevaluates organization, weapons, doctrine, logistics, C2 arrangements, C4I systems, and so on in a synthetic environment. At the end of the operation, forces come back and give, to the experiment team, real-world data or feedback to improve experimental conditions. In this scientific loop, the theory has been developed by the empirical data by using a method of inverse deduction.

A simulation is simply an executable model. The selection or development of the simulation(s) intended to support a transformation-related experiment is one of the most important activities in the preexperiment phase. The artificial world being simulated consists of only those components of the real world and those interactions among those components that are represented in data or in algorithms. A good model will include both the essential elements and adequate representations of the real world and real-world behavior or effects. It is possible to use computer-assisted exercises and to test three types of variables as follows:

- *People*—this includes the numbers, skills, and ways in which they are organized for operations.
- *Process*—how members of the military organization, coalition, or interagency team accomplish their responsibilities.
- *Infrastructure*—the systems, federations of systems, technologies, and other material resources that are applied in support of military operations.

The execution of the experiment could be with involvement of the different exercise teams or without the teams, using multiple repetitions of the same situation. Collection and analysis of data need to be synchronized with the design of the experiment (subject of research, hypothesis, and research aim) and the experimentation plan.

One advantage of model-based experiments is that they are potentially more flexible than experiments that involve human subjects (particularly large military organizations). So, activities normally associated with the postexperiment phase of an experiment can and should stimulate rapid iteration and evolution of the model, the experiment's design, and expectations [ALB02].

6.3 DATA ANALYSIS, REPORTING, AND PRESENTATION

After the execution of experiment, it is time to analyze the obtained results. The experimental team validates the results of experiment by using more than five measurement methods and by comparing qualitative observations with quantitative statistical results. When a quantitative method is used by experimenters, it is important to respect the fundamental statistical rules. For neutralization of possible mistakes, it is necessary to observe and compare the minimum of 30 basic statistical units in multivariate analysis, by not more than 40 independent variables, in order to provide an effective and precise graphic presentation of the final findings. When the experimental team uses qualitative methods, then it is necessary to have five or more expert opinions for statistically valid results. Embracing whole thinking about proper methodology and experimental tools is an important part of designing the experiment before execution; the team must be aware about experiment limitations. The confirmation or rejection of the hypothesis is optimized when the comparative analysis among qualitative observations and quantitative interpretations of findings are equal or similar. If somebody has doubts about the experiment findings, he or she should use same data and same methodology to confirm their opposite thinking.[5]

Concerning validation of the experiment model, Schellenberger [SCH74] has written that one way of organizing validity-related thinking is to recognize three kinds of model validity: technical, operational (tactical), and dynamic. Technical validity has four primary components, as follows:

- *Model validity*—refers to the model's correspondence to the real world (fidelity, or in the language above, this includes some aspects of construct validity).
- *Data validity*—refers to the validity of both raw and structured data. Raw data validity refers to the capacity to capture the phenomenon being modeled correctly or the accuracy, impartiality, and the ability to generalize. Structured data validity deals with abstract data, such as aggregated units.
- *Logical validity*—refers to the way in which model behaviors and results are considered.
- *Predictive validity*—do the model's results make reasonable predictions of outcome conditions? This is very difficult to confirm for models and simulations of future concepts and technologies. However, if a model is generating counter-intuitive predictions or its results seem inconsistent, the modeler will want to investigate the causal mechanisms with the model.

[5] [PoPo2] p.81.

In operational validation, the team should explore the outputs of the model and trace the relationships in the model from that output back to the changes in inputs that caused the change. Additionally, operationally oriented team members should be assessing operational reasonableness (face validity) of the force behaviors implied by the modeling. Dynamic validation explores the limiting conditions under which the model is valid. Two facets should be considered. The first is with respect to the insight it provides into the behavior of a force or concept. That is, experiments should not draw conclusions lightly about the performance of some platform or small-unit doctrine changes from a simulation whose entity resolution is at the level of flight groups and battalions. Second, experiment teams need to explore and understand the limiting conditions of the decisions that the model should reasonably support.

All of this thinking is basically knowledge collection for the future experimentation methodology and planning. Now, the experiment team should prepare the final report and presentation of the experiment findings. The final report must contain the following information:

- Stake–holders
- Topic and general hypothesis of experiment
- Subject, aim, and experiment objectives
- Management plan and timeline
- Experiment manager and experiment plan
- Participants
- Description of used methodology
- Short description of experiment chronology
- Reasons for eventual differences from experimentation plan
- Assessment of expenditure
- Level of reached objectives
- Analytical results
- Conclusions about research problem
- Proposals for implementation of experiment findings (for example, in armed forces)

The above-mentioned elements ensure a comprehensive report to the person or sponsor who placed the order for the experiment. The main purpose of the document is to provide an independent, objective, and responsible expert interpretation of the experiment findings. For the proper public presentation of report, it is necessary to know in advance the type of audience, their interests, and their background. The presentation needs to visualize the experimentation problem, design, execution, challenges, and final findings with appropriate and graphics. After the correct presentation of the experimentation process, the implementation of the results in the real world begins.

6.4 REVIEW QUESTIONS

6.1 What are the main steps of the experimentation process?

6.2 What are the functional elements of the experimentation plan?

6.3 Describe a model-based experimentation.

6.4 What is the structure of the experiment findings presentation?

PART II

COMBAT MODELING, COMPUTER-ASSISTED EXERCISES, AND PRACTICE

7

COMPUTER-ASSISTED EXERCISE (CAX) ARCHITECTURES

7.1 DISTRIBUTED EXERCISES AND DISTRIBUTED SIMULATION

Although the main technical characteristics of distributed exercises and distributed simulation had been explained in Chapter 5, this chapter will demonstrate the application of distributed interactive simulation (DIS) and high-level architecture (HLA) standards on multinational, multilevel, and multiresolution CAX.

Thus, DIS defines an infrastructure for linking simulations of various types at multiple locations to create realistic and complex virtual worlds for the simulation of highly interactive activities. This infrastructure brings together systems built for separate purposes, technologies from different eras, products from various vendors, and platforms from various services, and it permits them to interoperate.

The basic architecture concepts of DIS are as follows:

- **No central computer controls the entire distributed exercise**. DIS uses a distributed simulation approach in which the responsibility for simulating the state of each entity rests with separate simulation applications residing in host computers connected via a network. As new host computers are added to the network, each new host computer brings its own resources.

Computer Assisted Exercises and Training: A Reference Guide, by Erdal Çayirci and Dusan Marincic

- **Autonomous simulation applications are responsible for maintaining the state of one or more simulation entities**. As the user operates controls in the simulated or actual equipment, the simulation is responsible for modeling the resulting actions of the entity using a simulation model. That simulation is responsible for sending messages to others, as necessary, to inform them of any observable actions. All simulations are responsible for interpreting and responding to messages of interest from other simulations and maintaining a model of the state of entities represented in the simulation exercise. Simulations may also maintain a model of the state of the environment and nondynamic entities, such as bridges and buildings, that may be intact or destroyed.
- **A standard protocol is used for communicating ground truth data**. Each simulation application communicates the state of the entity it controls/measures (location, orientation, velocity, articulated parts position, etc.) to other simulations on the network. The receiving simulation is responsible for receiving the entity state information and for calculating whether the entity represented by the sending simulation is detectable by visual or electronic means. This perceived state of the entity is then displayed to the user as required by the individual simulation.
- **Changes in the state of an entity are communicated by its controlling simulation application**.
- **Perception of events or other entities is determined by the receiving application**.
- **Dead reckoning algorithms are used to reduce communications processing**.

A method of position/orientation estimation, which is called dead reckoning, is used to limit the rate at which simulations must issue state updates for an entity. The dead reckoning model represents the view of that entity by other simulation applications on the network and is an extrapolation of position and orientation state using a specified dead reckoning algorithm. Simulations are not required to report the status of their entities as often.

The management of a distributed exercise is desirable to facilitate the operation of the network and certain aspects of the distributed exercise. DIS management functions can be divided into network management and simulation management. Network management functions handle the basic network functions, such as load management, monitoring of nodes and gateways, and error recovery. A network manager would also have knowledge of host computers on the network, which includes their physical locations and network addresses. The network manager would perform an analysis of network performance. Management of the overall distributed exercise is also needed. The functions of exercise management include starting, restarting, pausing, and stopping of an exercise; creating and removing entities from an exercise; and collecting and distributing data with simulation applications. This standard includes protocols for the best effort and reliable communication of these exercise management messages [MCF04].

Interaction between simulations is achieved by the broadcasting of protocol data units (PDUs), each of which contains a packet of data that describes an event in the simulation. To reduce network traffic, entity states in a DIS simulation are dead reckoned. PDUs are broadcast using the unreliable user datagram protocol (UDP)/Internet protocol (IP), so PDUs may be lost. However, the effect of lost PDUs is minimal because of the use of dead reckoning. Each PDU contains a time stamp to indicate when the PDU was issued. Entity PDUs contain dead reckoning information that other simulations use to keep track of where the entity is located. The simulation that owns an entity uses both a high-fidelity model and a dead reckoning model. When the difference between the two models exceeds a threshold, the simulation issues an updated entity PDU. The command to start or resume a DIS simulation contains both the simulation time and the actual time at which the simulation is to start. DIS simulation time is the reference time (e.g., universal coordinated time) within a simulation exercise. Simulation time is established ahead of time by the simulation management function and is common to all participants in a particular exercise. The major functional components of the HLA are federates, the run-time infrastructure (RTI), and the interface between the federates and the RTI. Federates are the simulations that participate in an HLA federation. The RTI acts as a distributed operating system for the federation. Two services that the RTI provides are object data exchanges between federates and time management. The HLA enables federates to manage local time in a way that allows them to coordinate data exchange with other members of a federation. The HLA time-management structure supports interoperability among federates using different internal time-management mechanisms. The HLA supports these capabilities provided that federates adhere to certain requirements necessary to realize each service.

To achieve these goals, a single, unifying approach to time management provides time-management interoperability among disparate federates. Different categories of simulations are special cases in this unified structure, and they typically use only a subset of the RTI's full capability. Federates need not explicitly indicate to the RTI the time-flow mechanism (time stepped, event driven, independent time advance) being used within the federate but must use the RTI services (including time management) that are appropriate for coordination of data exchange with other federates. Data sent from one federate to another can be delivered by the RTI either in delivered order or in time-stamp order (TSO). For the TSO, the sending federate has to time stamp the outgoing data and the receiving federate has to be able to receive the TSO data. The RTI ensures that the TSO data are delivered in TSO. In general, distributed simulations are simpler than the mixed federations that HLA can support. Consequently, the time management for a distributed simulation can be much simpler, using a single approach instead of the many that the HLA has to support [BAK07]. Distributed exercise is providing a more flexible, technologically achievable training method, which has been used lately for the bilateral (Figure 7.1) and multilateral (Figure 7.2) cooperation of military forces and other security elements of countries.

Example 7.1

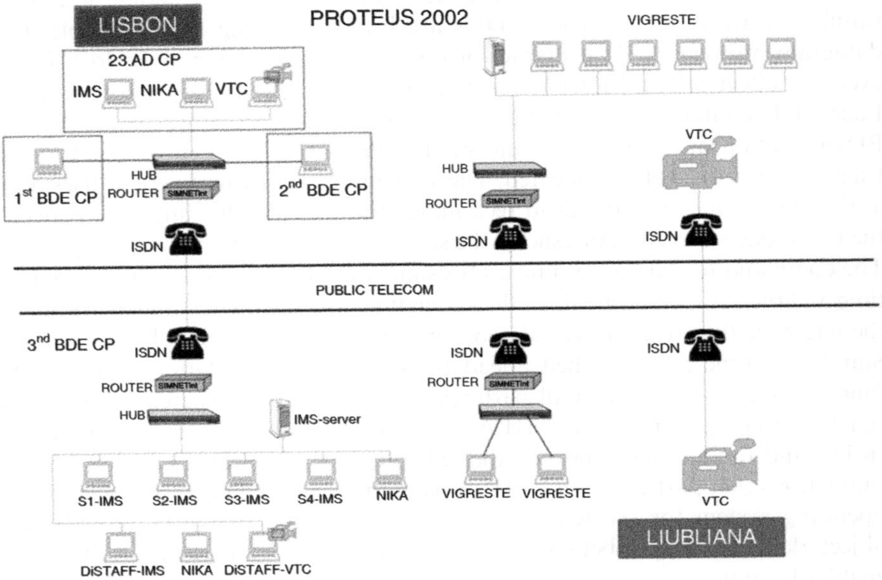

Figure 7.1: Distributed CAX architecture among Slovenia and Portugal PROTEUS 2002.

During Distributed CAX Proteus 2002, military staff colleges formed three bilateral brigade head quarters (HQ) and Divisional HQ as higher level commands (HICON) simulation exercise was combat one with some elements of civil military cooperation (CIMIC). The Portuguese simulation center distributed combat simulation model VIGRESTE, whereas the Slovenian simulation center distributed an information management system as an intraexercise C4I program, which was used for communication, exchange of COP, reports and returns, observations, and analysis of training audience efficiency. Thus, for these two countries, that approach was technologically and conceptually achievable. The main aim of the CAX was to enhance bilateral cooperation and to exchange experiences and knowledge for managing this type of exercises.

The multinational distributed exercise SEESIM 04 took place in 2004 under an initiative called the South-eastern Simulation Network and was used to strengthen the cooperation between countries on Southeastern Europe (Balkans). The main aim of the exercise was to improve the exchange of information among national security systems of participating states. The simulation model joint theater level simulation (JTLS) was distributed to other countries from the Turkey simulation center in Istanbul, where exercise control (EXCON) was located with all the functional elements

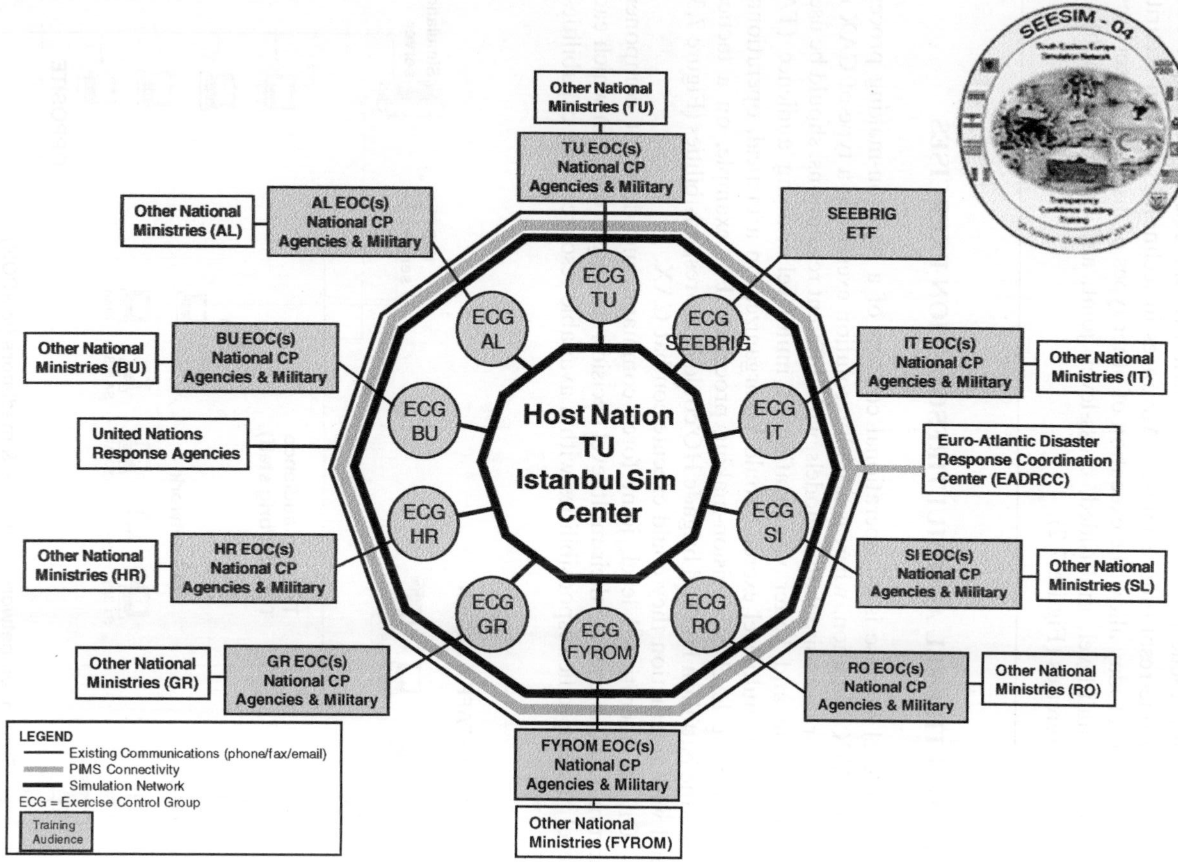

Figure 7.2: Multinational distributed exercise in Southeastern Europe SEESIM 2004.

and deployable elements on each remote site, which was structured by military and civilian security elements. The exercise was primarily dedicated to examining possible disaster situations and the development of common search and rescue procedures. Both distributed exercises were simple one-level and one-resolution exercises. Advances in technology and in security procedures today allow the execution of other types of distributed exercises, such as multilevel, multisided, cross-level, joint, and combined (multinational) exercises (Figure 7.2).

7.2 MULTILEVEL AND MULTIRESOLUTION EXERCISES

A multilevel exercise is an operational concept of a decision-making process during CAX execution, whereas a multiresolution exercise is a type of CAX in which two or more simulation models, with different resolutions, should be used for execution and better visualization of multilevel training audience (TA) decisions. A multilevel exercise could be organized on a tactical, operational, and strategic level of decision-making processes. For example, on a tactical level, the battalion HQ and brigade HQ could share responsibilities (Figure 7.3), and with interaction they could execute combat CAX.

On an operational level, joint force command and different component commands can execute their multilevel decision-making process, in which each level has a specific responsibilities within available resources and capabilities.

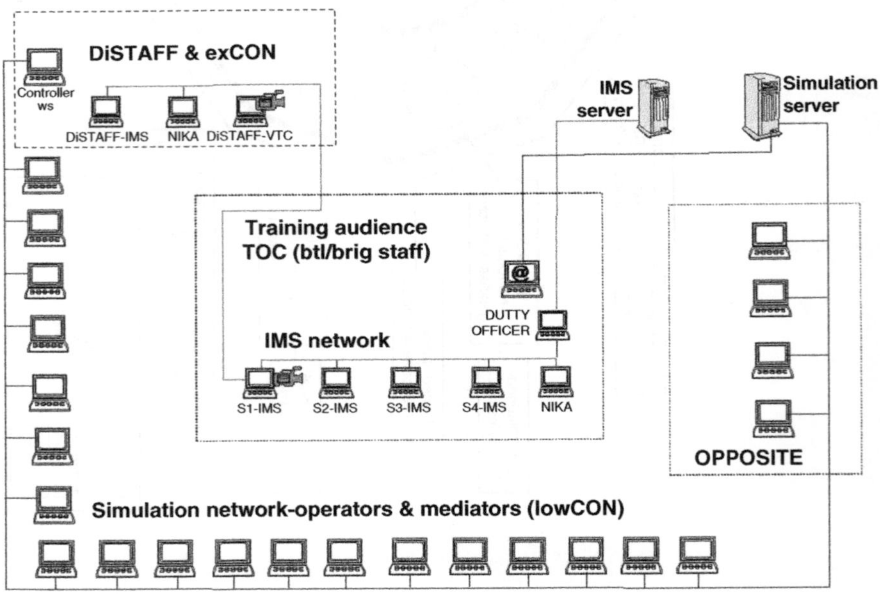

Figure 7.3: Multilevel CAX on a tactical level.

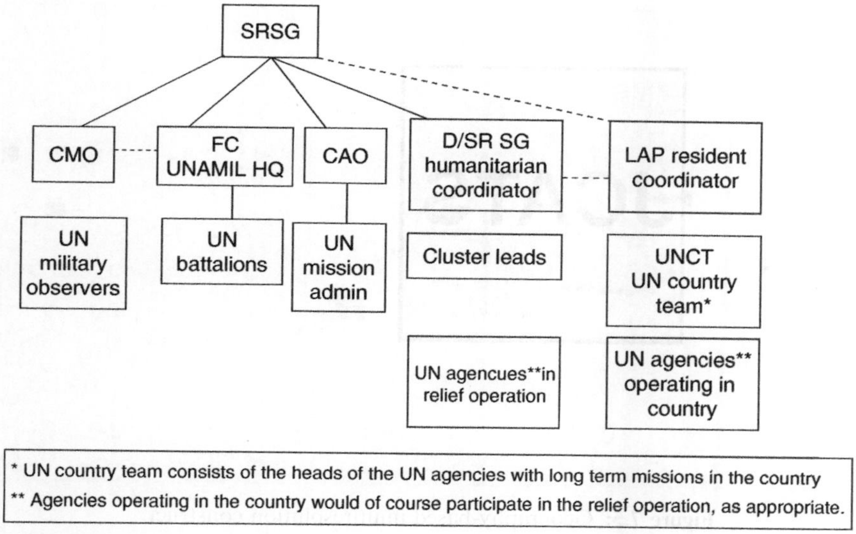

Figure 7.4: Complex multilevel and multisided entity structure in the CRO CAX environment.

Contemporary operations have a multisided character that has been reflected also in CAX for a crisis response operation (CRO), where civilian and police elements, local authorities, and international organization (IO)/ nongovernment organizations (NGOs) are becoming a part of the overall exercise structure. So the complexity from real-world operations has already been a fact in recent CAX, where coordination and cooperation among all involved sides are the main training objective and challenge (Figure 7.4).

Multiresolution exercises will be used in the future for better visualization of decisions on a tactical level, and at the same time, the operational level HQ could proceed with their activities on a lower resolution. As an example, Cayirci [CAY07] explained the new project for the North Atlantic Treaty Organization (NATO), which will provide an initial operating capability of multiresolution distributed modeling and simulation, in the form of a NATO Training Federation (NTF). NTF will be based initially on joint conflict and tactical simulation (JCATS) and JTLS. JTLS is a joint, highly aggregated, constructive simulation system, which fits operational or higher theater level simulation requirements. In JTLS, the terrain is modeled by hexagons. A typical hexagon size is between 6.5 and 7.5 km from one end to the other (i.e., one side is between 3 and 3.5 km). That best fits when the simulated units (simulation entities) are battalions or wings/air packages, that is, multiple aircrafts in an air mission, and ships (frigates, submarines, etc.). JCATS is a joint, high-resolution constructive simulation, where the details like a single troop can be simulated by using high-resolution terrain and environmental data. It is also possible to aggregate the simulated entities into units and command them as aggregated units in

Figure 7.5: Geography-based multiresolution construct.

JCATS. When the training audience is lower than or equal to component command (corps) level, JCATS provides better fidelity simulation. NTF will connect these two constructive simulations through HLA (i.e., RTI). In NTF, the entities simulated in JCATS can interact with the units in JTLS. For example, an aircraft in JCATS can fire a missile to a unit in JTLS. When this capability is available, highly aggregated JTLS can be used as long as higher resolution simulation is not needed. When the simulation resolution is higher than the one that JTLS can provide or the simulation of high-resolution tactics is required, JCATS can be used. The outputs of these two simulations update the attributes of the units and entities in both simulations (Figure 7.5).

The conventional exercise structure and process need to be modified for multiresolution exercises because both tactical and operational level training audiences with different training objectives will be present at such an exercise. This has an impact on almost everything, including scenarios, databases, map and geographical data requirements, EXCON organization, and communication and information system (CIS) environment.

In multiresolution exercises, there will be multiple sets of training objectives. Moreover, each set of training objectives will require events and incidents designed in resolutions different from each other. Although there are different sets of objectives and multiple resolutions for events and incidents, all single white truth of them should be coherent and consistent. Similar to simulations, a highly aggregated event may encapsulate several high-resolution events. Or an incident in highly aggregated level can be perceived as an event at the high-resolution level. This requires very careful coordination of events and incidents between different resolution levels. Moreover, the planning made for a highly aggregated incident may create many incidents not planned and not expected for the high-resolution TA. All these incidents make an impact on the scenario, which needs to be enough for all sets of training objectives, as well as resolution levels [CAY07].

This new concept it is at the same time a CAX and an experimentation in preparation, organization, and execution of the multilevel exercise. With the effective preparation phase, which is creating conditions for execution, all involved participants can gain enormous benefits out of simulated process. Also, the multilevel TA can execute their decision-making process on separate but interdependent parallel levels of operation.

7.3 CROSS-LEVEL, JOINT, AND COMBINED EXERCISES

In 2000, the Revolutions in Military Affairs (RMA) emerged as a modern strategic approach toward operations; it exposed global changes in the understanding of the military force's role in the security environment. Strategic-level publications emphasized terms such as *system of systems, dominant knowledge, and global reach*; all of them explained the need to respect the main security dimensions in the area of operations and not just an opposite military forces. Along with its obvious technological implications, it was important to accept the notion that "combined with the dramatic changes in military doctrine and operational concepts (the RMA) fundamentally alters the character and conduct of military operations."

Embedded in the critical attributes of the strategy are the following attributes:

Jointness. Identify and strengthen those specific capabilities that enable the contingency force (CF) to fulfill Canadian security priorities, deliver a joint capability to deal with weapons of mass destruction, information operations, and other asymmetric threats; and form counter-threat partnerships with domestic and international partners.

Command and control. Foster jointness in command and control, as well as logistics and intelligence, which includes the development of a deployable joint headquarters that can exercise national command and logistic support of main contingency forces.

On the operational level, battles and campaigns are conducted to achieve strategic objectives. Military doctrine holds that it is essential to maintain a direct link among the strategic through the operational to the tactical level. Battles must be fought and campaigns must be waged for strategic ends. There must be a direct relationship between the deliberations by political decision makers and the weapon pit or foxhole. The political and diplomatic aspects of any alliance should support and be directly supported by operations in the field and fleet [LEA98]. And that statement is valid for most states in the international community. With the multinational willingness and participation in operations such as the wars in Afghanistan and Iraq, the nations express the need for a combined, multinational answer on asymmetric threats to the global security. All above mentioned trends influence training systems, procedures, and technology.

With the respect of modern trends in training, CAX methodology is like avant-garde for training of all involved military, civil, and police forces in the area of CRO. During the joint CAX on the international level, organizers usually using NATO combined joint task force (CJTF) concept, which consist of the following TA HQ: joint force command (JFC), land component command (LCC), maritime component command (MCC), air component command (ACC) and Supreme HQ. The cross-level interaction is usually triggered by interactive injection into the exercise process on component level (tactical level). Staff reactions on injection are steering coordination among different boards on component level, which is shaping decisions and in same time supporting joint coordination process on operational level among joint coordination boards. Finally in specific circumstances an operational conditions demands reaction on strategic level. One of the examples is partnership for peace (PfP) Simulation Network, in particular the peace support operation (PSO) CAX serial VIKING, where the basic principle is to use modern available developed tools and methods in a multinational environment to enhance aim and objectives, mainly related to joint staff procedures (cross-level interaction) and civil-military cooperation (Figure 7.6).

Example 7.2

The CAX VIKING 05 was conducted from December 5 to 16, 2005, with more than 2000 participants representing 25 different NATO and PFP

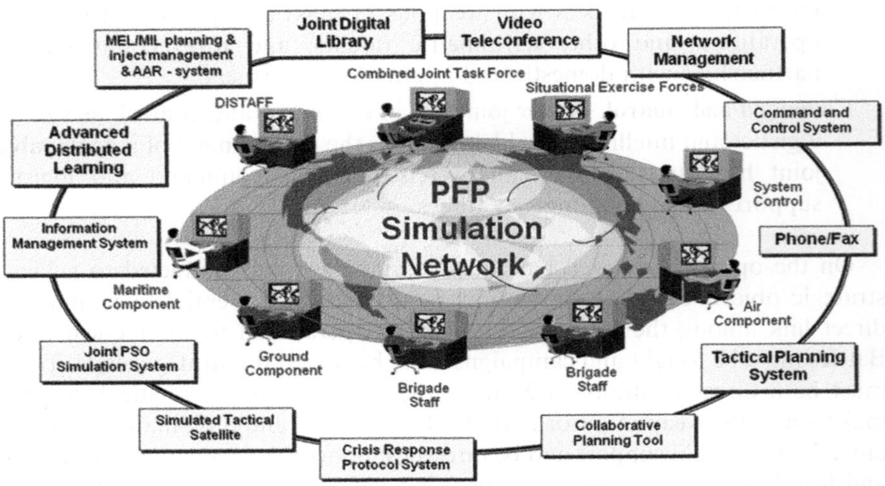

Figure 7.6: PFP simulation network.

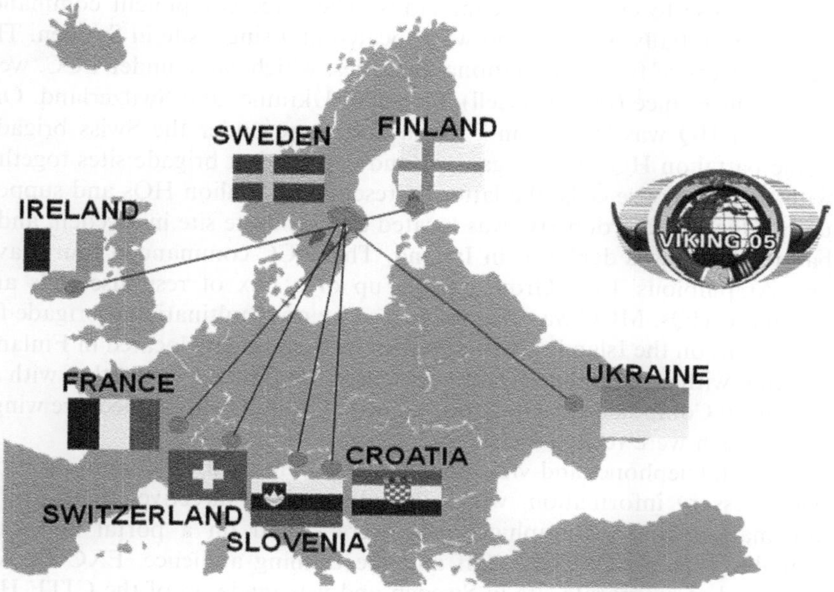

Figure 7.7: CAX VIKING 05 EXCON and TA locations.

countries. Viking 05 was a peace support operation exercise in the spirit of the partnership for peace, reflecting a peace enforcement operation under a UN Security Council Resolution in a fictitious country named Bogaland. The exercise control organization and training audience were located at 11 sites in eight different countries (Figure 7.7). A tunneled Internet network linked the command control system and the necessary exercise control systems. NATO standards and procedures were used. Sweden was the lead nation with outstanding support provided by both the United States Joint Forces Command (USJFCOM), based on a bilateral MOU, and the NATO.

The main aim of the exercise was to train and prepare individuals for participation in units and HQs using the NATO CJTF concept. The training objectives included planning at operational and tactical level and the use of standing operating procedures (SOPs) within a CJTF concept as well as the practicing of CIMIC and civil military relations (CMRs) in a CRO. The training organization was also used for experiments aimed at extending the development within the framework of the partnership for peace simulation network.

The civil component was made up of about 200 officials and included the senior representative of the UN Secretary General (SRSG) and his HQ. International and national organizations were represented at all sites and made injects and acted as role players. A nucleus CJTF HQ was organized as higher command within EXCON, mainly to give a realistic interface with and response to component commands. The three component commands were substantially manned and were located at a single site in Sweden. The headquarters of four multinational brigades, which came under LCC, were located in France (response cell), Slovenia, Ukraine, and Switzerland. One battalion HQ was located in Croatia and came under the Swiss brigade. Some battalion HQs were organized and deployed at brigade sites together with their response cells, the latter representing battalion HQs and support units. One battle group HQ was located at a separate site in Sweden, and a battalion HQ was deployed in Ireland. The MCC commanded four Naval and Amphibious Task Groups made up of a mix of response cells and organized HQs. MCC was also reinforced by one multinational brigade for operations on the Island of Gotland with its exercise HQ located in Finland, together with a Battalion HQ and response cells. The ACC together with an organized Combined Air Operation Center (CAOC) commanded five wings, all of which were response cells.

E-mail, telephone, and videoconferencing were used for communication. All necessary information, which includes orders, directives, background information, and geographical data were stored on a portal and were available to both EXCON staff and the training audience. EXCON was organized at a separate site in Sweden and was made up of the CJTF HQ, SRSG with HQ, and joint exercise control JEC, which coordinate all injects and role-play gaming activities. Site exercise control (SEC) units, which report directly to the JEC, controlled all response cells deployed at the component site and at every brigade, task group, and battalion site. The EXCON organization used a simulation system (Tyr) to provide a common situational picture and also to simulate the effects of situation development and orders given to respond to normal frictions, terrain conditions, and actions by belligerents and the local population. Gaming was built on a Story Board and was documented and tracked in the "Exonaut" computer system, which was available to all EXCON staff [ISB05].

This was typical CAX where the cross-level interactions in joint and combined environment were achieved by DIS and a distributed exercise, with approximately 300 injects for different decision-making levels, including civil and police organizations.

The exercise analysis showed that TA achieved the main aim of the exercise: To train and prepare individuals for participation in units and HQs using the NATO CJTF concept. All the major training objectives were met for the conduct of joint operations using NATO standards and procedures, incorporating the civilian dimension, in both the planning and execution phases and for the further development of the VIKING distributed and computer-assisted exercise concept. Exercise Participants have therefore enhanced their ability at all levels to contribute to a future NATO CJTF-led operation. The TA has increased local capability for peace in the simulated area of complex emergency for 70% on 74 municipalities of Bogaland (Figure 7.8).

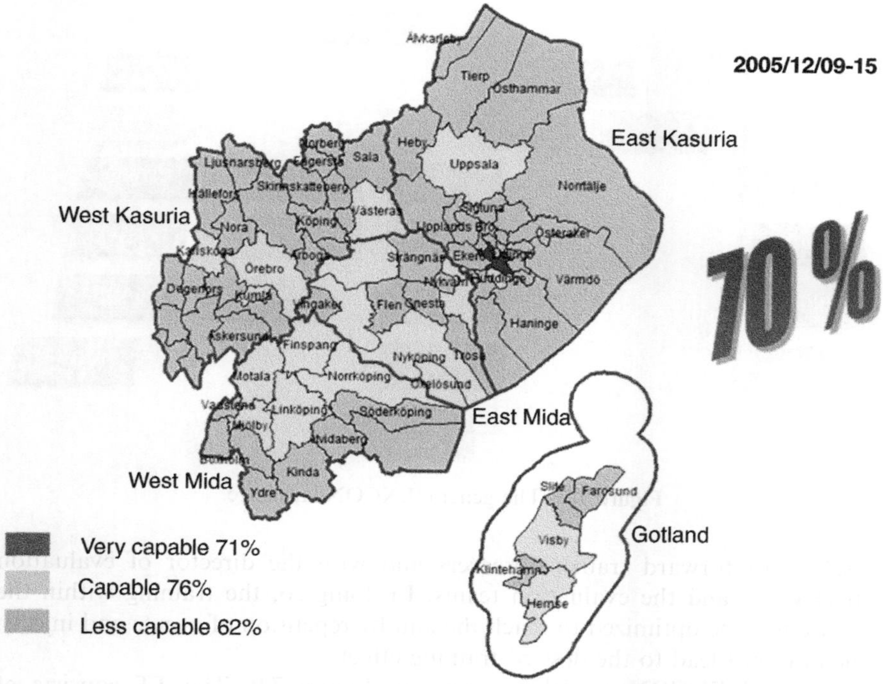

Figure 7.8: Efficiency of TA on CAX VIKING 05.

7.4 EXCON STRUCTURE

The EXCON structure should provide training conditions and deliver training operational content. It has both direction and control functions that allow it to establish the conditions necessary for achievement of the exercise aim and objectives and training objectives (TOs) by the TA. As the chief of EXCON (Figure 7.9), the exercise director (EXDIR) may steer exercise play—in both direction and tempo—as necessary to enhance learning opportunities, reinforce key lessons, and achieve objectives. He or she also has the authority to terminate, modify, or suspend an exercise or parts thereof if concerns for safety of participants or others dictate. In the event that the TA is established in dispersed locations, it may also be necessary for EXCON to disperse its elements to provide the most suitable coverage, assistance, and control. These elements normally consist of, at minimum, a liaison officer (LO) but may also include training teams (TT), senior mentors, role players, administrative support and umpires. Reliable communication means among EXCON elements are critical for their effectiveness, and these may include telephones, radios, e-mail and video-teleconferences. The exercise control Staff shall work

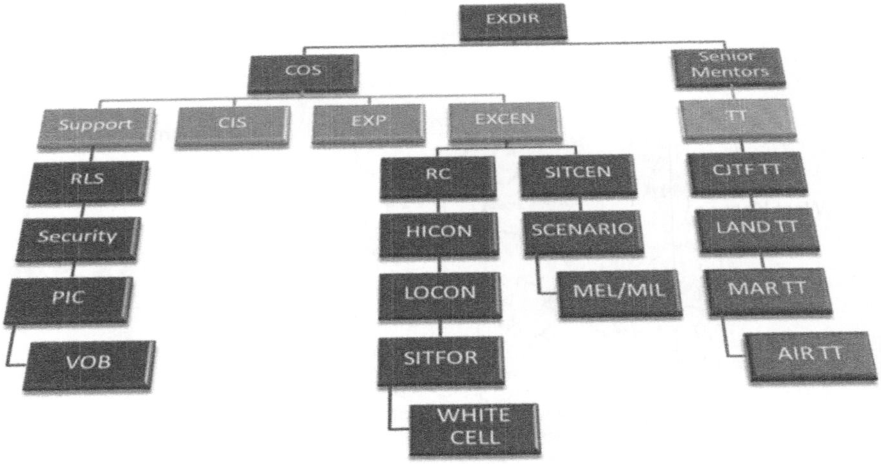

Figure 7.9: The generic EXCON structure.

closely with forward trainer-observers and with the director of evaluation (DIREVAL) and the evaluation teams. In doing so, the training within the exercise may be optimized to reach the aim by repetition of events and injects, which did not lead to the desired training effect.

A typical EXCON model is shown in Figure 7.9. The TT consists of mentors, observer/trainers (O/T), subject-matter experts (SME), and analysts. The TT is deployed with TA, observes TA, provides onsite instructions and training, and collects input for the after action review (AAR) and the evaluation of TA. The exercise center (EXCEN) is the organization responsible for the consistent and coherent flow of the exercise according to the exercise objectives. The experimentation team runs the experiments planned in conjunction with the exercise. Finally, the support team has the elements like real life support (RLS), visitor officer bureau (VOB), public information center (PIC), security office, and C4 event support team.

EXCEN functions can be categorized into five broad classes as the situation center (SITCEN), simulated higher level command (HICON), lower level commands (LOCON), white cell, and situation forces (SITFOR). SITCEN monitors the current status of the exercise closely and steers it according to the training objectives. HICON and LOCON represent the command levels/echelons that would normally be at the level above and below the TA, respectively. The white cell is a response cell that is composed of SMEs or role players that represent agencies, organizations, institutions, and individuals outside of the NATO or the opposing force structure. SITFOR is the cell that manages the status of all the NATO and opposing forces in the scenario except for the ones represented by TA, HICON, and LOCON [CAY07].

Cayirci [CAY07] explains that the first important team in SITCEN is the scenario team, which is responsible to maintain and manage the scenario related information, and to respond to the request for information (RFI). They have a close connection and coordination with the SITFOR cell. The master events list/master incidents list (MEL/MIL) team is another important team within SITCEN. The MEL/MIL manager closely monitors the status of every injection and manages the dynamic scripting and injection of incidents. The MEL/MIL team ensures that every injection serves for a TO, and the coverage of TA, TO, and the exercise duration by the injections are well balanced.

Event managers manage the main events from start to end. They ensure that the TOs for the event are achieved by the TA. They closely follow the feedback from O/Ts and response cells (RCs); design new incidents consistent with the scenario, simulation, the overall event and TO; and ensure that the planned incidents are injected at the correct time (i.e., consistent with the scenario and status), to the targeted TA. They coordinate with the SMEs in functional desks, CAX coordinator, and MEL/MIL manager. The CAX support team in the SITCEN provides the interface between the CAX delivery team and SITCEN.

7.5 RESPONSE CELLS

RCs represent subordinate, higher, or flanking units and formations, plus other agencies and characters that may operate within, around, or even beyond the theater of operations. In the event that one of the main task force elements, such as a component, is not participating as a TA, then that headquarters should be represented as an RC. It is best that RCs are staffed by officers actually from those units, formations, and agencies. Examples of the latter would be IO/NGO, local governmental organizations, and media of all types. RCs provide the continual, direct, and realistic interface with the TA through injects simulation play, reports and returns, press releases, media stories, telephone calls, radio calls, and so forth. RCs representing units and formations subordinate to the TA are known as LOCON. The RC representing higher or flanking units, formations, and headquarters are known as HICON. For exercises with a TA at the operational level (such as a joint force command HQ), HICON will have to portray strategic political and military organizations. They may have to provide role players for key individuals such as force commander (FC), UN Secretary General (UNSG), UNSG Special Representative, and so on.

LOCON and HICON consist of RCs. The number of RCs is dependent on the scenario and the TA. Each RC is made up of a MEL/MIL coordinator, several planners, a battle captain, and many simulation operators. Planners in an RC act as the subordinate or superior of the TA. When an order or report is received from the TA, planners make their plans according to the current situation and the doctrine. Then, they hand over the plan to the battle captain. The battle captain is the interface between the planners and the simulation operators. He/she is an experienced operational planner that also has an insight about the capabilities of the simulation. He/she converts the plans to a sequence of orders for the simulation. The operators than enter the orders into the simulation. The results and reports from the simulation are also passed to the planners, who transform them into realistic reports for the TA. Standard reporting procedures and formats should be applied in passing the results to the TA [CAY07].

The white cell (WC) represents agencies, organizations, institutions, and individuals outside of the NATO force structure and outside of the opposing forces structure. The composition of the WC must be tailored for each exercise, but typically it may include IO/NGO, local governments at federal/provincial/municipal levels, local police forces and local civilians.

7.6 TRAINING AUDIENCE

The TA has the main role in execution of the CAX. Because of usage of distributed exercise and DIS, most of the TA has been exercising on their peacetime locations. During the operational planning process, they are preparing the concept of operation (CONOPS) and operational plan (OPLAN) that culminate with Joint Coordination Order (JCO) before start of exercise (STARTEX). The TA should be individually and collectively ready for the CAX in accordance to scenario and their JCO. More conditions for achievement of their training objectives will be provided by EXCON. During the execution of the CAX, the main task of the TA is the decision-making process on different levels and production of exercise documentation, with the solutions for the injected incidents on CAX. At the time of the exercise execution, the TA should follow and respect doctrinal and operational rules and regulations. Exercise specification documents define what type of exercise will be executed (joint, combined, multilevel, multiservice, etc.) and on which level is the primary TA.

Interactions among TA boards, operational centers, planning teams, and assessment teams at the time of decision-making process and their respective products represent the main aim of training on CAX. The entire battle rhythm on CAX is connected with the theoretical origins of the decision-making process and its practical application. The decisions quality is displayed by simulation model in interaction with terrain and operational entities data bases, and it is integrated with the remarks of observers' team on the CAX.

A professional approach toward crises management on CAX means the creation and common understanding of all kinetic and nonkinetic activities until the desirable success has occurred.

7.7 REVIEW QUESTIONS

7.1 What the DIS defines?

7.2 Make a list of the basic architecture concepts of DIS.

7.3 Are there any differences between distributed exercise and distributed simulation?

7.4 What is the purpose of the multiresolution exercise?

7.5 Why is it necessary to organize a cross-level exercise?

7.6 Describe the vital positions in EXCON structure.

8

CAX PROCESS

Knowing and understanding computer-assisted exercise (CAX) architecture by accepted distributed interactive simulation (DIS) and high level architecture (HLA) standards is just a first important step in the preparation of requested training conditions for training audience. The next step is to develop a concept for CAX processes in terms of planning, preparation, execution, and analysis.

8.1 EXERCISE SPECIFICATION

Military education is defined as the permanent process of the examination of staff, units, and force components with regard to their capabilities to conduct a given mission and task effectively and efficiently. It also investigates the validity and reliability of systems, procedures, programs, and objectives. It includes the analysis, assessment, feedback, and lessons learned. To set up proper training and education conditions, the military authority has to define purposse, objectives,[1] and exercise methods. Military exercises prepare commands and forces for operations in times of peace, crisis, and conflict. Therefore, the aims and objectives of military exercises must mirror the current and anticipated operational requirements and priorities. The same applies for CAX, which is one of the modern educational methodologies or exercises for the military forces. CAX could be organized on strategic, operational, and tactical level; or

[1] A desired goal expressed as the performance under set conditions for a defined standard. It describes the knowledge, skills, or attitudes to be reached during training.

Computer Assisted Exercises and Training: A Reference Guide, by Erdal Çayirci and Dusan Marincic

on a joint or component level, with the involvement of military, civilian, and police structure operation to execute multicultural, multinational, and multi-sided crisis-response (CRO) simulated exercise.

The exercise specification (EXSPEC) is the superior military command order [in this context we will name that body as an officer scheduling the exercise (OSE)] to the officer conducting the exercise (OCE) to plan, conduct, and analyze the exercise. The EXSPEC is developed as a collaborative effort with the OCE and with other headquarters (HQs), agencies, and centers participating in the exercise. The EXSPEC should include the geostrategic situation and the draft exercise milestone planning schedule. It shall provide enough details to support the development of the exercise plan (EXPLAN) but allow for the shaping of the exercise by the OCE to ensure full achievement of the exercise aim and objectives. It is essential that the final draft EXSPEC be promulgated prior to OCE initiating exercise planning and development activities (nominally 12 months before exercise execution). The EXSPEC is drafted by the exercise planning group (EPG) based on the OSE's decisions, direction, and guidance, and it is coordinated with the HQs involved and finalized at the exercise specification conference (ESC). The steps supporting this activity are as follows:

- *Draft Initial EXSPEC.* The initial draft of the EXSPEC is prepared by the EPG and distributed within the OSE HQ and to the OCE, the participating evaluation team, the TA, the participating nations and the officer directing the exercise (ODE), when designated.
- *Develop Module 1—Geostrategic Situation.* The geostrategic situation module is prepared by the EPG and should include a generic description of the crisis area, the major regional actors, a summary description of the current crisis, and the historical background of the crisis, as well as the major political, military, economic, cultural, humanitarian, and legal conditions that support a military response. The geostrategic situation is summarized in the EXSPEC main body and expanded in an EXSPEC Annex.
- *Develop the Draft Exercise Milestone Planning Schedule.* The draft exercise milestone planning schedule should include the major exercise events to be included within the EXSPEC. It will be updated throughout the exercise process.
- *Determine and Coordinate Exercise Budget Responsibilities.* The OSE Fund Manager must coordinate responsibilities with the OCE fund manager and the ODE when designated.
- *Determine the Exercise Public Information (PI) Policy.* The OSE will determine the PI policy for the exercise.
- *Conduct Exercise Reconnaissance, Liaison, and Coordination Visits.* The OSE officer of primary responsibility (OPR) will determine the requirements for; assign tasks; and coordinate reconnaissance, liaison, and coordination visits conducted by or under the responsibility of the EPG. Host nation confirmation will be requested, and if applicable, site surveys will be conducted.

- *Determine Real Life Support Responsibilities.* The responsibilities for real life support should be clearly delineated in the EXSPEC to the ensure availability of resources sufficient to support achievement of the OSE's aim and objectives.
- *Coordinate Exercise Specification.* The EPG will facilitate coordination and incorporate appropriate comments on the OSE's Initial Draft EXSPEC for presentation at the ESC.

The ESC is convened by the OSE with representation of the OCE, TA, the participating evaluation teams, and the host nation(s) involved. The ESC's aim is to present the coordinated draft EXSPEC for endorsement. The main sections of EXSPEC document are as follows:

- Overall requirements should outline the requirements for the exercise, the purpose, and the relationships to other exercises.
- Exercise aim and objectives should be specified by the OSE. The objectives should be achievable within the scope and scale of the exercise. Additionally, mention of any evaluation and/or analysis requirements should be made here.
- Geostrategic situation should summarize the major regional actors, a description of the current crisis with the historical background of the crisis, as well as the major political, military, economic, cultural, humanitarian, and legal conditions that support a military response, including membership in relevant arms control treaties and agreements. An expansion of the geostrategic situation will be included as an annex to assist the EPG in determining scenario requirements.
- Concept of the exercise means how it will be designed to achieve the aim and objectives. The setting must be outlined, including the current and projected political, military, economic, humanitarian situations, and so on, which provide the background for the development of the exercise planning documents [strategic military assessment (SMA), initiating directive (ID), strategic planning guidance (SPG), etc.], which will be used by the exercise players to initiate the operational planning process (OPP) for the exercise.
- Participating requirements—when deemed necessary, the intended employment of forces nominated to simulate certain events or military situations should be stated. Participation requirements should indicate the parties involved, envisaged kind and size of headquarters, envisaged number of troops, types of armed forces involved, envisaged levels of command participation, any other important exercise information and in the case of a synthetic exercise (SYNEX), the extent of simulation should be indicated.
- External support requirements to be requested from other HQs and/or agencies. For example, support from communications and information system (CIS), staff augmentation, special analysis/evaluation teams, or international/nongovernmental organizations.

- The possible political implications of the exercise should identify areas of possible difficulty and the recommendation on implications with respect to arms control treaty/arrangement membership (i.e., the Conventional Aimed forces in Europe (CFE) Treaty and Vienna Document 1999 provisions).
- Planning schedule should include the proposed exercise process (EP) schedule. It is necessary to include the dates by which firm allocations and nominations of forces are required.
- The public information policy for military exercises should be transparent to the public.
- Visitors and observers section should include the details for organizing a Visitors' and Observers Bureau (VOB), Visitors and Observers Days and any Distinguished Visitor Day (DV-Day) proposed for the exercise.
- Administration, logistics, real life support, and funding requirements should include all requirements that cannot be met within current OCE capabilities. This includes the estimated cost for the exercise.
- Reporting requirements should include the necessary guidelines on the form and timelines for reporting on the exercise. This section will also include the need for a postexercise debrief.
- Other special instructions include other necessary guidance, for example on how the exercise design will be conducted, conditions for initiation of the crisis response planning, etc.

8.2 PLANNING AND PREPARATION

Qualitative and timely planning and preparation of the exercise are origins for execution of all kinds of exercises in particular for CAX. Planning has to take in account logic of planning procedure; the following actions must be performed by exercise planners:

- Study and analysis of the concept and conditions for the exercise execution
- Definition of purpose, exercise, and training objectives
- Determination of exercise participants
- Production of EXPLAN
- Planning of logistic and financial support

Although the planning is a separate activity of planning teams, it has to be synchronized with the four exercise training phases.[2] The EP consists of following four stages: exercise concept and specification development, exercise

[2] Phase I—Academics; Phase II—Operational planning process (OPP); Phase III—Execution; and Phase IV—After action review.

planning and product development, exercise operational conduct, and exercise analysis and reporting.

The planning and product development stage must provide all deliverables that enable the exercise to be executed successfully. The key deliverables are as follows:

- OCE's exercise planning guidance
- The EXPLAN covering all exercise phases and subphases of the operational conduct stage.
- Planning situation/scenario modules, which include documents and data covering the following:
  - Module 1—Geostrategic situation (part of EXSPEC)
  - Module 2—Theatre of operations information (at the beginning of planning process)
  - Module 3—Strategic initiation (before Phase I)
  - Module 4—Crisis response planning information (before Phase II)
  - Module 5—Force activation and deployment information (between Phase II and Phase III)
  - Module 6—Execution information (start exercise package for Phase III)

The exercise planning and product development stage is organized under the 11 key activities with specific steps for each activity. The first step is activation of core planning team (CPT) from organizers and the training audience side. Core planning teams represent and coordinate the interests of all participating HQs in the exercise. During the planning process, they discuss organizational and conceptual issues on three main CPT meetings, which have been conducted in line with the exercise planning conferences. The second step is the analysis of OCE requirements and limitations to establish a proper organizational framework. The next step is for the OCE to issue an exercise planning guidance, which explains the aim, scope, principal objectives, funding, and delineation of responsibilities for the exercise planning process. After that, the CPT starts to develop an initial draft EXPLAN. The scenario group has responsibility to develop and produce Module 2—Theatre of operations. Then a site survey and initial planning conference (IPC) are conducted. The IPC planning activities have been managed by different functional syndicates, which have the obligation to prepare initial plans to support execution of the exercise. For example, plans for application of computer, information systems, and forms of civil-military cooperation; exercise design; scenario; exercise locations; real-life support; logistic support; and other necessary exercise plans. The steering committee has an arbitrary role to solve all kinds of challenging issues. CPT develops a refined draft EXPLAN and products. The next important milestone in the planning process is the main planning conference (MPC). A working methodology consists of syndicate discussions, where participants from all exercising HQ exchange their views on developments of organizational and conceptual preparations for the execution. Usually, the MPC is an opportunity to finalize most supporting exercise plans and planning products.

After MPC, the EXPLAN and products should be finalized. The final coordination conference (FCC) is dedicated to the final confirmation of readiness for exercise execution from an organizational, conceptual, and documentation point of view. This conference is the last opportunity for the training audience to express additional interests and proposals for influencing the exercise design before the start of the CAX. Following FCC, The Chief master events list/ master incidents list (MEL/MIL) is organizing with the event managers, scenario group, and CAX simulation experts at the MEL/MIL incident development conference and the MEL/MIL scripting workshop. Both events are dedicated to the development of MEL/MIL, which represents future developments of the exercise scenario and provides appropriate conditions for achieving training objectives during execution of the CAX. The STARTEX validation is basically the transfer of exercise incidents to the simulation model and the synchronization of CAX data bases with a desirable chain of exercise incidents.

8.2.1 Scenario Development

The Exercise scenario provides conditions for computer simulation modeling of natural and societal processes in a holistic way. The entire process should ensure the timely recognition of societal deviations, correct procedures for their stabilization, and objective analysis of affected society. CAX ensures that training audiences can observe the effects of their own decisions and can improve or adjust the exercise decision-making process to the circumstances of the artificially affected society.

Proper computer simulation of a contemporary operation demands a common understanding of the international security environment, characteristics of the area of complex emergency, and peace force procedures in rebuilding the local society in the area of responsibility. Modern national security is a part of wider international environment, where security, beside states and their alliances, is more and more in the domain of a global security system. The contemporary security is in wider sense multidimensional and consists of political, economic, demographic, social, and environmental dimensions. International peace operations are a functional part of both national and international security planning, which means inevitable cooperation with the structural elements of UN is required. Thus, exercise scenarios need to cover all the above-mentioned dimensions and the interactions among them.

The international community has to perceive, in timely manner, the critical circumstances in certain region, to understand the root causes of the conflict and to decide and plan the intervention with the peace forces as a security instrument. Recent crises in the world suggest that the UN is still the main security provider for the crisis areas. The successful management of the UN operations demands an effective decision-making process of UN main bodies such as the Security Council, Secretary General, Department for Peace

Operations, Department for Political Affairs, and Department for Humanitarian Affairs. These departments provide the requested conditions for components of peace forces to fulfill their obligations in the area of operations. Besides the military component, police component, and component of civil organizations, the media is also an instrument for peace enhancement, which sends clear messages to the population and mediates during societal conflicts. In last decade, the media acted as a supervisor on international agencies and NGOs activities in areas of complex emergency.

All the above-mentioned elements of the international community create a dynamic of nonlinear processes on perceived critical areas, with both qualitative and quantitative analytical methods. The Exercise should explain social-economical unevenness among social groups, violation of political and human rights, demographical pressures, legitimacy of political authorities, distrust among ethnical groups, absence of independent media, and other societal issues. The most important condition for the CAX is to use the security information about the affected society in the physical environment with the geographical location. The geographical conditions for execution of the operation have a crucial impact on peace forces' preparation and also on the establishment of a common understanding on the history, habits, mentality, and culture of the affected local societies.

With the respect of previously described information the origins for peace operation scenario in an artificial environment are in place. The scenario provides the initial conditions in simulation model, which requires a transfer of all stakeholders in the area of peace operation, to synthetic environment, where we can simulate and control interactions among them. Theoretical development of a scenario is not focused on '*what will happen*', but rather '*what are the necessary and sufficient conditions for a given result to be obtained?*' In practical terms this means, that the CAX is more focused on the outcome of the coordination processes, which leads to the desirable end state in the area of operation. The natural progression from the initial exercise scenario is MEL/MIL development process.

8.2.2 MEL/MIL Development

The MEL/MIL development process has been established because of the technological underdevelopment of simulation models. Most models are primarily focused on cold war, pure combat warfare, or kinetic engagements. But the real life does not care about simulation model weaknesses. Contemporary operations deal with the society in its entirety, and in particular with the kinetic warfare. Thus, the MEL/MIL has been used as a tool for covering automatic event generator gap, which uses the scenario as its basis. The responsibility that MEL/MIL is providing proper training conditions for TA, in accordance with approved training objectives, lies on chief MEL/MIL as a manager of the event managers (EM) coordination group. He is responsible to manage a MEL/MIL strategy meeting to decide about proper training events (e.g., kinetic, nonkinetic, asymmetric threat, and humanitarian assistance). The Chief also directs the MEL/MIL incident development meeting, where all participating sides discuss desirable incidents under each exercise event. The last event is MEL/MIL scripting workshop, which is dedicated to the creation of suitable injects to support each incident.

The MEL/MIL should include all injects provided to the TA, from the initiating directive that starts the crisis response planning phase, the situation updates prior to the command post exercise (CPX), and through to the end of the exercise (ENDEX). Thus, the key requirements of those drafting the MEL/MIL will be the approved EXSPEC, the scenario outline, and the approved training objectives. The scenario and MEL/MIL groups will usually hold their meetings during the exercise planning and product development stage.

The EM has to show ownership for their events and to assume full responsibility for the development of storyboard and delivery of the exercise. The exercise injects have to be created for the decision-making level of the TA. On tactical level, injects need to trigger decision-making processes for battalion, brigade, division, and core-level headquarters. On an operational level, injects have to activate joint coordination process on joint force command (JFC) level and on component level [land component command (LCC), maritime component command (MCC), air component command (ACC), special operations component command (SOCC), and psychological operations component command (POCC)]. The most important is to provide a flow of exercise in the spirit of jointness. The chief MEL/MIL and the EM should take care about the quality and intensity of injects throughout of the exercise execution. Balance between exercise requirements and battle rhythm must to be shaped after the STARTEX validation conference among the TA, database management team (DMT), scenario group, and EM. The prepared MEL/MIL is just a part of static scripting prior STARTEX; during execution, the TA reacts to certain injects. If the reactions are not in line with the expected outcome, then the chief MEL/MIL, scenario representative, EM, and response cells (RCs) have to create dynamic scripting to achieve a desirable effects on TA. As a result, achievement of the exercise and fulfillment of the training objectives (TOs) must be fulfilled.

8.2.3 CAX Databases and Database Management Process

Combat models and combat model data are explained in detail in Chapter 9. Combat models are generally fixed, which means that they are not modified for each exercise. However, combat model data needs to be populated, then carefully verified and validated for each exercise. This major effort requires the involvement of both technical and operational staff from various parties.

Data categories in a typical constructive simulation system are depicted in Table 8.1. It is not always necessary to change the data in all of these categories for every exercise. Model data are rarely modified for exercises. Environment data is also typically fixed, as long as the geography for the exercises does not change. Prototype data are relatively dynamic. However, the change in

Table 8.1: Combat model data.

Category	Sub–category	Example data items for sub–category
Model data	Modeling parameters	– Random number seeds and distributions
		– Model-related parameters
	Lethality data	– Point hit and kill probabilities
		– Area kill probabilities
		– Lanchester coefficients
Environment data	Terrain data	– Terrain features
		– Elevation data
		– Obstacles, buildings, etc.
Prototype data	Prototype data	– Unit types
		– System, (i.e., equipment and combat systems and types)
Exercise data	Unit and system data	– Individual systems
		– Individual units
		– Command hierarchy
		– Logistics relations
	STARTEX data	– Initial unit and system locations
		– Initial unit and system status
		– Unit and system arrival times
	External event data	– Types and times for external events
	Weather data	– Weather fronts
		– Weather conditions
		– Sun set and rise times

prototype data is generally limited. The most dynamic data are the exercise data, which need to be populated almost from scratch for every exercise. For this purpose, a data base team that consists of the representatives from the following parties is constituted ideally right after MPC:

- Database team leader: The team leader is actively involved in building the data base. He is a technical person that has also a strong operational background. This person accumulates the most of knowledge and experience in a CAX support team. He/she also has a key role for the management of a CAX during execution phase.
- Data base builders: The data base team leader can be supported by one or at most two technical database builders. The number of database builders should be limited.
- Data providers from the training audience: The exercise data for the training audience should be provided by the training audience. The representatives from all major components in a training audience should collect data related to their part of order of battle (ORBAT) and ensure that those data are correctly entered into the database.

- Data providers for situation forces (SITFOR) including operation forces (OPFOR): These are usually representatives from scenario team. Except for ORBAT related to the training audience, data for every other units, systems, and entities present in the scenario are provided by this group.

The data base team manages the data base building process in the following stages. Please note that this is a recommended generic process. Some stages can be merged with another or skipped for some exercises.

- Data base team constitution: During IPC CAX, the syndicate leader announces the required personnel for the data base team. The team comes together normally during MPC and is informed about the data base process.
- Data base definition meeting: Soon after MPC, the data base team convenes. The team members are provided with a better insight about the model data base, and the model, environment, and prototype data that will be used in the exercise are introduced. Members are also trained about the tools for collecting the data base.
- Data base review meeting: This meeting occurs typically four or five weeks after the data base definition meeting. The data collected by the team members are reviewed by data base builders together with data collectors. Missing data are identified and either asked to be provided as soon as possible or completed during the meeting.
- Data base validation meeting: Within 2 or 3 weeks after the data base review meeting, the collected data are merged into a single data base. Then, the merged data are uploaded into the model and verified, and any errors are corrected. Finally, the simulation is run, and the behavior of the model observed based on a carefully designed test plan. If the model behavior is not accurate, then the related data items are corrected.
- Start of exercise (STARTEX) validation meeting: As soon as STARTEX data are available, they are entered into the system. Ideally at least 3 or 4 weeks before STARTEX, STARTEX positions, and status of units and systems are examined by the training audience and scenario builders. Note that training audience representatives do not observe the data related to OPFOR. After this point, the data base is frozen.
- (C2) data base initialization: After the STARTEX validation meeting, the simulation data are downloaded into the formats required by C2 systems, and C2 databases are uploaded with the data in the simulation system.

8.3 EXECUTION

The exercise process begins with Phase I—Academics, where the training team provides requested lectures on procedures for participation of the military forces in CROs and on the OPP. Phase II—OPP follows the theoretical preparation of the TA. During the mission analysis of a certain area of complex emergency, the TA performs operational planning process activities to create a concept of operation (CONOP) and operational plan (OPLAN). OPLAN is important for the Phase III—Execution, because it offers content for a joint coordination order (JCO) or coordination directives for TA during CAX. For Phase III, the CAX has to have a proper exercise functional elements or structure. The TA is the center for all activities, where the exercise director (EXDIR) and exercise control (EXCON) provide requested exercise conditions. EXCON consists of an exercise center (EXCEN), SITFOR, RCs white cells (WCs), and training teams (TT), which deliver the exercise and at the same time control the outcome of the TA decision-making processes. The EXDIR has a significant freedom, to use EXCON structures that are most appropriate for achieving the OSE's exercise aim and objectives, as well as the OCE's TOs. Therefore, his *directs* the EXCON who *controls* the exercise execution to set the conditions to allow the OCE to achieve the exercise aim and objectives established by the OSE. The TA will conduct activities in accordance with the appropriate policies, doctrine, and processes as well as directives and their standard operating procedures (SOPs). So CAX has main elements (Figure 8.1), which help to activate the TA to fulfill TOs, which include: scenario, injects as a tactical problem, model, simulation, and analysis.

As was described previously computer simulation support, with its hardware and software, is essential for the preparation, execution, and analysis of the exercise. During execution, the simulation model presents interaction among entities, individuals, and structure in a synthetic environment. At the same time, it ensures data stream for communications (text, picture, and sound) between the TA and EXCON in real time (Figure 8.2). The generic execution process of CAX starts with the preplanned inject from injector (EM, RC, WC), which is received by the TA at the joint operation center (JOC). The JOC disseminates information from the inject to the appropriate functional board, for short-or for long-term planning. It depends on required actions. The designated functional area groups decide about required actions.

At the end of the day, the main joint coordination board (JCB) issues directions for action to a certain operational component or element. A serial of injects should create an operational pattern for the decision-making process and provide enough information for the TA campaign as well as an operational

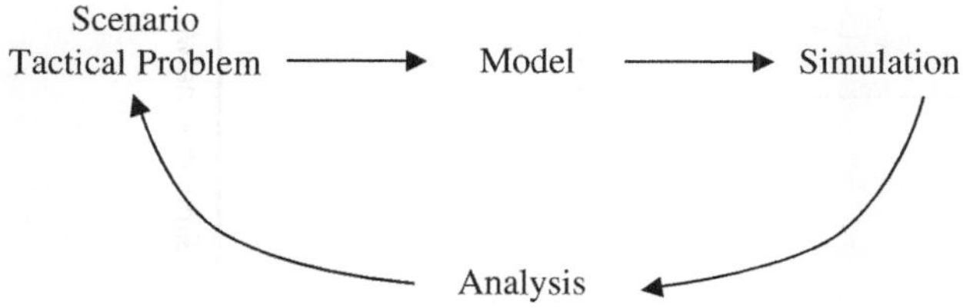

Figure 8.1: Elements of CAX.

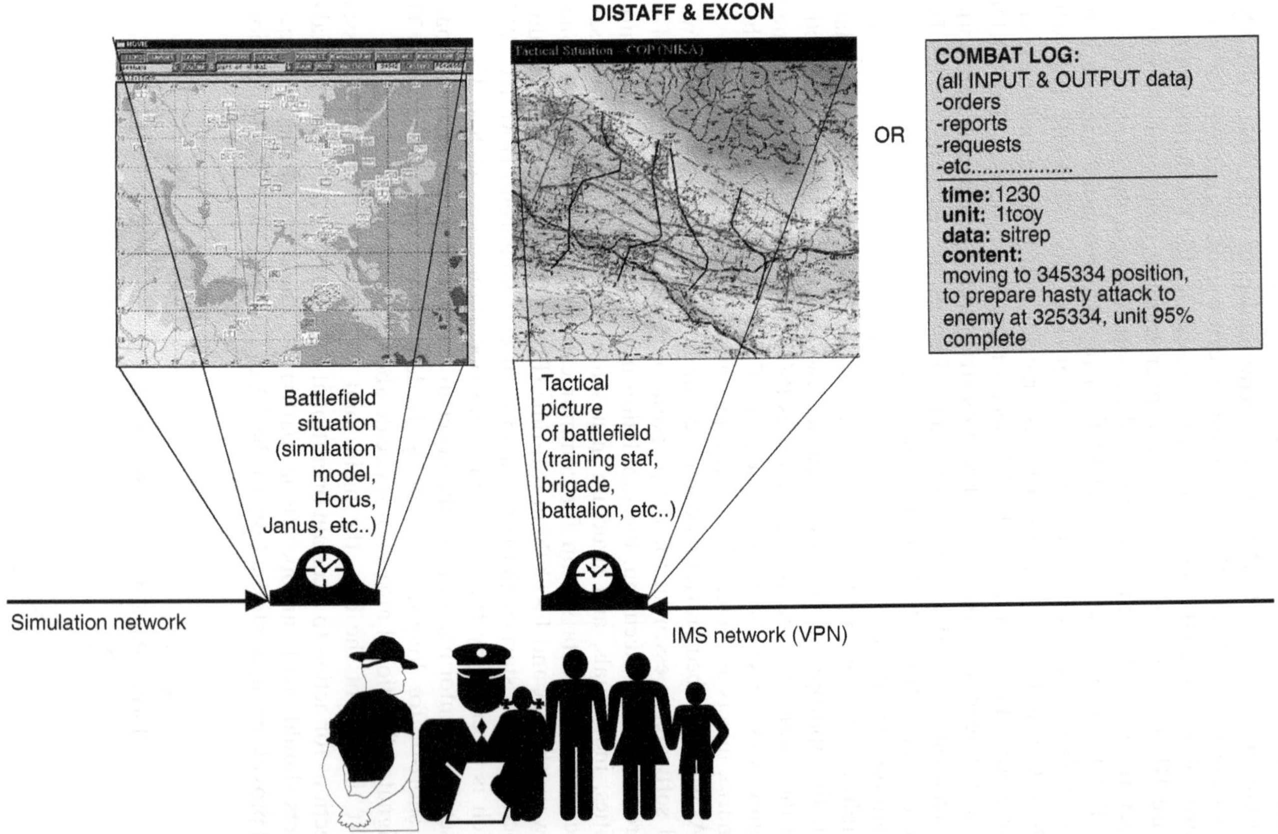

Figure 8.2: Controlling TA activities on the CAX.

and tactical assessment of the situation. Training teams observe joint coordination processes on the remote site, and together with the EXCEN (Figure 8.2) observations, the teams provide a concise picture to the EXDIR, which includes happened and what needs to be modified to steer the CAX toward desirable TOs. Most of the explained activities should be visible in the EXCEN.

For the analytical purposes, the simulation model saves all initiated activities and exercise data for the quantitative and qualitative analysis of TA reactions on different exercise injections. At the end, it is possible to run comparative analysis of exercise events, with the TA decisions in exercise time and space.

The primary goal of CAX is interaction among TAs during the decision-making process and their exercise products (JCO, FRAGO, analysis, etc.). The entire working process on CAX is connected to the theoretical origins of the decision-making process and its practical application. The simulation model presents the quality of TA decisions through the interaction among terrain data bases and operational components, and integration with the assessment of the EXCON.

The use of simulation models permits decision makers to determine nearly all the outcomes of the war game. As an excellent and cheap instrument of analysis, simulation enables the effects of plans, tactics, or doctrines to be tested in a variety of environments by repeating and replaying the scenario.

Although the entire training process occurs in a synthetic environment, it ensures improvement in the decision-making skills of TA, decreases expenditure, increases security for participants, and protects the natural environment in the long term.

With the application of CAX methodology in a training cycle for peace operations, there are new perspectives for joint training and preparations among civilian part of society, in terms of crisis management in the area of complex emergency. The professional approach toward crisis area means the common understanding of all kinetic and nonkinetic components in the decision-making process, and their interaction until the desirable end state [MAR02].

8.4 ANALYSIS

The exercise analysis is Phase IV in the overall exercise process, and it is the most important phase. The main task for this CAX activity is to provide to all CAX participants the observations and information in the form of an after action review (AAR) support package (Figure 8.3).

The exercise analysis begins with the collection of observations and conduct of evaluations as well as assessments compiled throughout the exercise process. It includes postexercise analysis and reporting by the TA to the OCE and supporting organizations in accordance with requirements and procedures established in the EXPLAN.

Three categories of data and information should be used to support exercise analysis and reporting as described below. The EXPLAN should set the

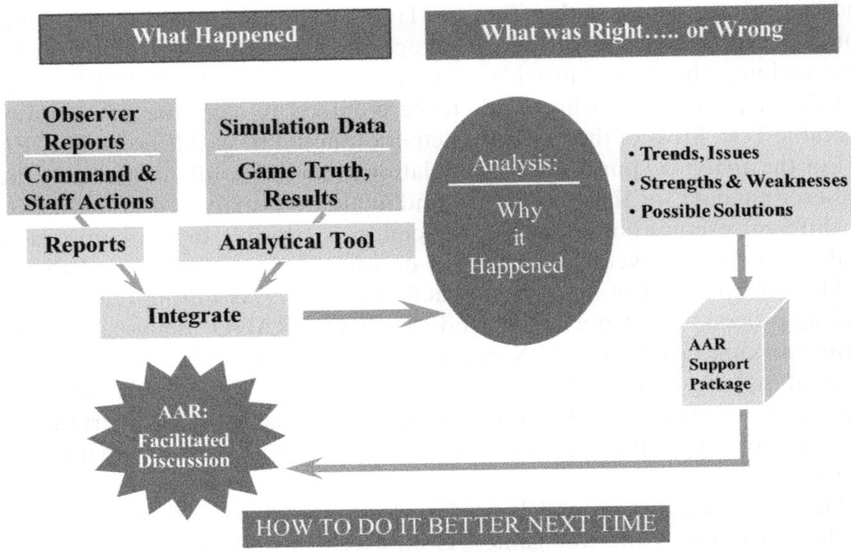

Figure 8.3: After action review process.

requirements, timings, and responsibilities for collection, archiving, and appropriate distribution of each category. There are two major categories of exercise analyses and reports — those that address the performance and accomplishments of the training audiences and those that address the planning and execution of the exercise, analyses of specific objectives, or experiment aims. TA data and information are perishable, and the requirements for their collection by designated means should be laid down in the EXPLAN.

Throughout all the training phases of the exercise, the TAs should be making, collecting, and processing observations in accordance with their command lessons learned program. The messages, decision briefings, video teleconference (VTC) tapes, records, and reports that are produced by the TAs during the course of the exercise should be archived for exercise analysis and reporting purposes.

These documents may also be examined by analysis and evaluation teams in pursuit of identification and justification of potential recommendations for improvement of doctrine, SOPs, and so on. If the exercise includes the crisis response planning phase, the TA HQs should collect and archive the HQs' official records, orders, information exchange reports, requests, command historical records, visual information, and combat documentation.

The command diaries and battle logs of events and decisions are an essential tool for the preparation of the TA first impressions reports (FIR). Data from command and control as well as from common operational picture systems should be periodically copied and archived for postexercise analysis and reporting purposes. The requirements for back up and archival of the operational data should be laid down in the TA HQs' OPLANs.

The EXPLAN should establish the specific requirements for EXCON staff together with evaluation, analysis, and experimentation teams to collect data and information as well as record their observations and comments throughout the entire exercise process. These data, information, observations, and comments will be in various forms as follows:

- EXCON logs of events, decisions, and daily meetings are an essential tool for exercise analysis and reporting. The EXCON log should assist those staffing the FIRs to put the comments into perspective as well as enhance the objectivity of the final exercise report (FER).
- A wide range of personnel can be tasked with observing the training audience or particular aspects of the exercise and its support functions. These observations must be collated by authorities detailed in the EXPLAN and their products passed to the OCE for analysis during the staffing of the FER.
- Where capabilities are marked for assessment during an exercise, the reports should be made available to the OCE before staffing of the FER.
- Training process reports, which provide information on how well the exercise delivered the TOs for the TA and supported the training needs of the TA, should be made available to the OCE for development of the FER.

The analytical products should be in quantitative and qualitative formats. Most analytical products about TA effectiveness have been produced by qualitative analytical methods. Only quantitative analysis for combat simulation has been produced in connection with existing data bases about combat entities. For example, the data include the number of losses, consumption of material, combat potential, and so on (Figure 8.4).

The observers and TA produce descriptive assessments on a daily basis, as well as an evaluation of the overall efficiency of decision-making process on CAX. For a proper overview on TA performance, it is advisable to combine these qualitative products with the quantitative measurements of the TA impact on security dimensions of the synthetic crisis area. A multivariate analysis of local capability for societal reconstruction from the real world crisis area could represent an initial situation in the synthetic environment. By measuring the changes of the potential of main security dimensions (demography, social situation, economy, political situation, and environment), the CAX analyst could have an assessment of the daily effectiveness of the TA decision-making process.

Each exercise inject has a capability index (Figure 8.5), which could influence the potential of security dimensions, if the TA decisions are in line with expected outcomes. Observations about the efficiency of TA decision-making processes after each inject allows a permanent examination of their effectiveness and fulfillment of training objectives on the CAX.

At the end of each exercise day, the daily results should be added to the analytical model. The total sum of all capability indexes at the ENDEX shows

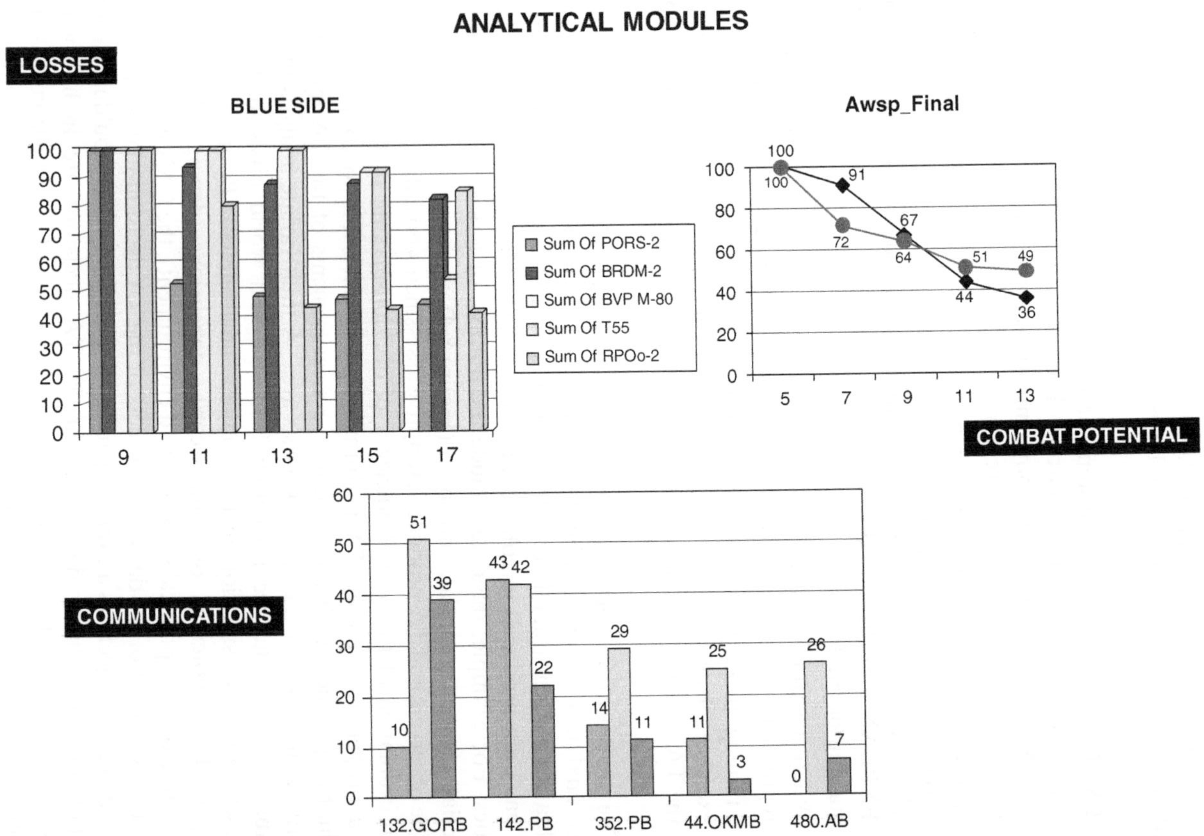

Figure 8.4: Example of analytical modules for combat CAX.

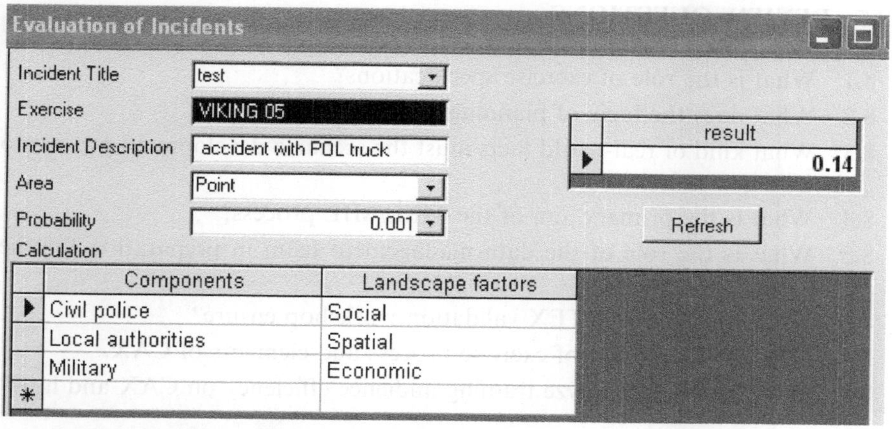

Figure 8.5: Calculation of capability index for exercise inject.

transparently to the CAX participants and EXDIR, also level of local capability for societal reconstruction in the area of CRO, and an assessment of the TA's efficiency in the given exercise time and space (Figure 8.6).

With the societal variables of security dimensions, it is possible to create a synthetic environment of the complex emergency on CAX; proper preparation and training of joint operational headquarters and their components is important to understand conditions on the real crises areas. Computer modeling of artificial societal conditions can lead to the extremes, which do not exist in the real world [MAR05].

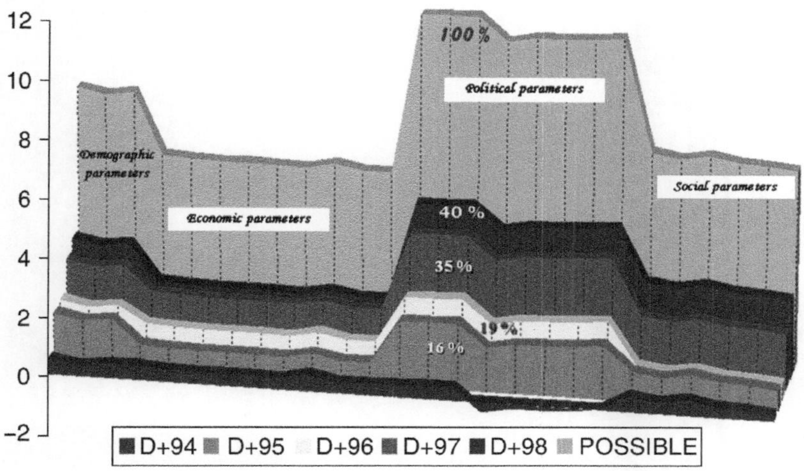

Figure 8.6: Analysis of TA's efficiency at ENDEX.

8.5 REVIEW QUESTIONS

8.1 What is the role of exercise specifications?

8.2 What does the logic of planning procedures look like?

8.3 What kind of real-world facts must the scenario development process to take into account?

8.4 What is the primary aim of the MEL/MIL process?

8.5 What is the role of the data management team in preparation for the CAX?

8.6 What should STARTEX validation workshop ensure?

8.7 Describe the phases of exercise process and elements of CAX.

8.8 Is it possible to analyze training audience efficiency on CAX and how?

9

COMBAT MODELING

Modeling is the creation of models, and it is the key for the systemic approach to the solution of a particular problem. It means the transfer of attributes and characteristics of a research problem from the real environment to a similar problem in a synthetic environment, but to a smaller extent and in a different form. Thus, a model is an approximate picture of the real world. It has to be simple and understandable but representative and complex enough to explain credibly the problem [HAG67]. For effective modeling, it is necessary to understand the contents of the problem and simultaneously the mathematical language or the method in which the problem has been described and modeled [SAV00].

Modeling allows the transition from the real world to a virtual one (artificial or synthetic), where attributes of landscape components, processes, and events have been transferred. We could represent their structure (static models) and functions (dynamic models) with the models [GEO67], and we can forecast future developments or a certain chain of events, which is a particular value of simulation models [VRI82].

Chorley [CHO71] has classified simulation models as scale models, analog models and mathematical models. Scale models represent the real world with the same material on a smaller scale (for example, presentation of river bank erosion or a military model for tactical actions); analog models use different materials. Mathematical and operational models have the highest level of abstraction, in which characteristics of events are presented with mathematical equations or other symbols [VRI82].

Computer Assisted Exercises and Training: A Reference Guide, by Erdal Çayirci and Dusan Marincic

The growing cost of operational trials and limited budgets makes extensive live tests of military systems and frequent live exercises nearly impossible. Defense planning deals with extremely complex military system behaviors and the necessity of using of innovative analysis tools. Simulation is one of the most employed methods for the military analyses, training, and acquisition. Its increasing importance is now well recognized in most of the armed forces. Dramatic advances have been made in military simulation community, but these advances are associated mainly with computer technologies.

Computer simulation modeling has been more described than defined in the literature. The main function of computer simulations is to investigate the characteristics and outcomes of the system, which is too complex for the logic and mathematical analysis [FIS02]. Nardi [NAR93] offers the following description: "Computer simulation modeling is ensuring scientific investigation with simplified analogy in order to have better analysis and understanding of some phenomena. It is focused on management of computer experiments, where the primary goal is behavior of the system in time, with the mathematical and logical operations. Computer simulation is powerful method, which is capable to process a great number of variables, and present complex system, with the simulation of effects of variables on many cycles."

Davies and O'Keefe [DAV89] have proposed the following: "When the word simulation has been used by computer scientists, statisticians or management scientists, they understand with that, creation of abstract model, which is representing certain real world system. Simulation is describing proper aspects of the system as a serial of interactions implemented in computer program. "Most simulation descriptions propose that simulation models adhere to the following guidelines:

- Have been used for modeling systems that are too complex for modeling with the ordinary analytical models
- Consists of logical or mathematical operations on variables
- Can represent many variables and operations
- Are holistically oriented toward modeling all aspects of the whole system
- Are focused on presentation of processes
- Are focused on research and experimentation

The differentiation of simulation models from the other models involves what we are doing with the model and not which type of the model we have. In the case of simulation, we want to understand the behavior of the model and ways to manipulate it. The value and purpose of the simulation is measured by the level of data application from the model [DYK81].

The epistemological problem of nonlinear modeling can be crudely summarized as a dichotomy between engineering and science. As long as a representation is effective for a task, an engineer does not care what it implies about the underlying mechanisms; to the scientist, however, the implication makes all the difference in the world. The engineer is certainly concerned with minimizing implementation cost, but the scientist presumes, at least, to be focused on what the model means vis-a-vis natural laws. The engineering view of science is that it is mere data compression; scientists seem to be motivated by more than this. Much of what constitutes the new science founded on chaos and complexity actually derives exactly from computer-based simulation exercises, which involve both the iconographic modeling of chaos/complex mathematical functions and the simulation of artificial ecosystems in a way that seems to demonstrate the possibility of emergent properties. In these abstract exercise which are an essential part of the chaos/complexity project, mathematics and programming come together in an experimental form.

Social simulation studies provide an opportunity to fill the gap between empirical research and theoretical work while avoiding the individualist tendency of most mathematically based approaches. In particular, social simulation provides not only testing hypotheses but also an observation of social processes. Combat modeling is also part of social simulations, because military is one of the three main pillars in society: political authorities, military, and civil society. However, we are examining combat modeling, which it is a part of social simulation processes.

It can, therefore, offer the basis of new efforts to devise categories of description and new analyses of social reality. In other words, social simulation can provide instruments for modeling society. Contemporary work with the simulations is a kind of theory development in which the simulation is understood as a formalization of theory. The question here is not "what has happened" or even "what might have happened" but rather "what are the sufficient conditions for a given result to be obtained?" While the first two questions are exclusively descriptive, the latter may have prescriptive consequences. It may provide hints about how to enhance or reinforce some social strategies by telling us, for example, under what conditions these strategies become stabilized [BYR97].

Cares [CAR04] has written that although great variety exists in the specific application of combat models, the fundamental structure of most combat models is one of two basic types: deterministic (closed-form) or stochastic (probability-based) combat models. A **deterministic model** is one that fully describes all states in a given system with a set of closed-form equations. For example, Newton's Second Law, $f = ma$, which defines force f as the product of mass m and acceleration a does not admit for randomness or uncertainty in the mathematical representation of force. Because it completely determines theoretical relationships in a closed-form representation, the Second Law equation is a deterministic model. Deterministic models were particularly useful before digital computers made automated calculations more routine for the analysis of large, intricate systems.

Starting in the 1970s, the widespread use of digital simulation facilitated the inclusion of randomness and uncertainty in analytical models. A model that explicitly includes randomness and uncertainty is called a **stochastic model**;

stochastic models are usually created by modifying one or more terms in a deterministic equation with random draws from some probability distribution. These models run a certain number of times so that some effects of randomness or uncertainty cancel out and model results converge to stable, aggregate statistical values.

Today's higher level combat simulations (e.g., those at division, corps, and theater levels) are best viewed as implementing aggregate state-space models. The basic notion is that the "state" of the system (the two opposed forces, their strategies, and the environment in which they fight) can be represented by a collection of variables such as counts of personnel and vehicles in an area and terrain factors characterizing that area, rather than the locations and current behaviors of all the individual entities, such as individual soldiers and tanks. Usually, the simulation then generates the predicted future state as a function of the current (aggregate) state. In more general formulations, there can be "memory effects" of previous states as well. Again, the variables that affect this prediction are not just the sides' strengths. Instead, the predicted change of state depends on many other factors, such as terrain, defender preparations, flank exposure, strategy, and tactics. One important change of state, which is typically made at the end of time periods or when some significant event occurs, is a change of strategy or tactics (e.g., a decision to attack or withdraw, or to maneuver reinforcements to a trouble area).

The basic structure of contemporary combat models is over 100 years old and is a direct product of science, technology, and philosophy. It is clear that the models, previously described, are inadequate for representing Information age warfare. To summarize, they are inadequate for the following three main reasons:

- A process that can aggregate and disaggregate fine-scaled behaviors by definition must treat these local behaviors as noise at the aggregate level. Such a process cannot adequately represent local arrangement of elements, clever use of information, or massed effects from distributed forces. These, of course, are all important Information Age warfare precepts.
- The models rely on mathematics to represent combat activities as independent processes. Networked processes are by definition interdependent.
- The distribution of networked performance is highly skewed as feedback loops create "tipping point" behaviors.

Although network-centric warfare (NCW) concepts are said to capitalize on this fact, contemporary models actually enforce regularity and depend on less-skewed performance distributions. As mentioned above, standard stochastic modeling techniques depend on statistical convergence in a moderate number of runs. Simulations that contain models with skewed distributions require an exhaustively large number of runs for statistical representation. An Information Age combat model will require a transformation in military modeling philosophy and must therefore address these challenges by properly representing complex local behaviors, explicitly representing interdependencies, and capturing the skewed distribution of networked performance. In addition, an Information Age combat model must capture both the attrition processes and the search and detection processes important to distributed, networked warfare. Such a model would be a bona fide transformation in combat modeling philosophy and constitute a true Information Age combat model [CAR04].

9.1 TERRAIN MODELING

The preparation of an appropriate simulation model is a demanding task. Landscape systems are very complex, and they have distinguished behavior as an opened systems. Thus, it is particularly important to understand the relations among landscape components. With the help of the structure and dynamics of a landscape system, it is possible to understand the cybernetic task of its components, and with that the reactions and behavior of the system in its entirety. The designated simulation model has to explain real cybernetic interactions between all relevant living areas in a certain region or landscape, and it has to have the character of the real working tool and general validity, which means it is transferable to the other regions [VES91]. Terrain modeling is a process, which provides an essential structural element of the simulation model (Figure 9.1) called the digital relief model (DRM). The transfer of a real environment to the DRM[1] is an activity of environmental planners. Environmental planning involves the assessment of the suitability of an environment for different activities of social interest [MAR79].

Environmental data are basic elements information of about space. Landscape analyses use environmental data as relevant information for planning purposes. Thus, it is necessary to collect, from the very beginning, precise data about the areas of interest for the simulation model.

Analyses of landscape distribution are an important source of knowledge about disposition of environmental events, interaction among them, and their structure in the space [MAR79]. Three distinguished types of environmental events or their distribution are as follows:

- *Pinpoint distribution*—events are located in discrete points in space.
- *Line data distribution*—event is represented by the straight line, curve, or line in any other form.
- *Discrete plane distribution*—event in form of plane on the map.

[1] Terrain can be reviewed in three different ways. A *digital relief model* is a complex digital presentation of high-altitude relief pinpoints, significant curves, and terrain pinpoints, and even geomorphology. A *digital terrain model* is only part of a digital relief model, in which heights, inclination lines, significant curves and terrain pinpoints are presented only. A *digital altitude model* is a layer of altitude pinpoints on certain area. The digital relief model is the most important and fundamental layer in application of geographic information systems (GIS) [KVA97].

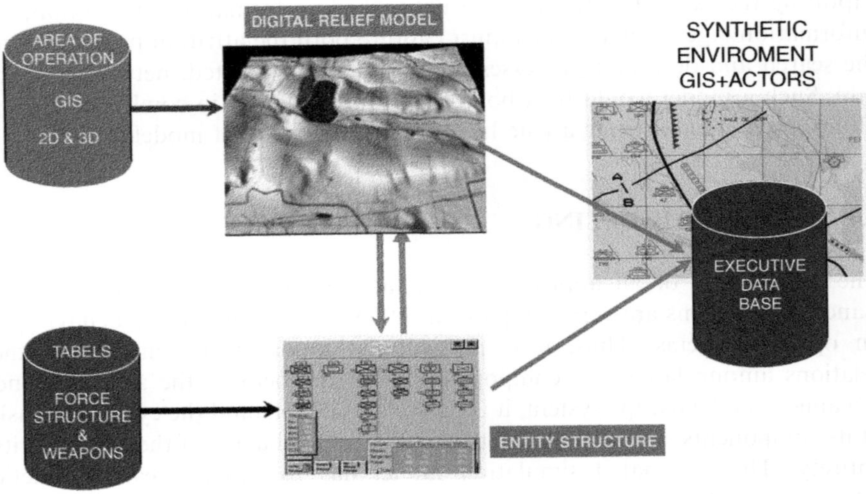

Figure 9.1: Structural elements of simulation model.

Formulation of the criteria is the first step toward creation of the model, which means creation from a conceptual perspective. The key elements of modeling are variables that influence the values of functional criteria [MAR79]. Taking in account suitability of space, the four distinguished groups of criteria are as follows:

- Criteria as result of the natural characteristics of space
- Criteria as result of anthropogenic characteristics of space
- Criteria as result of social and economical space
- Criteria of suitability as result of structural, systemic interdependence between development activities

9.1.1 Environmental Data Representation

A synthetic environment has been defined by the Defense Modeling and Simulation Office (DMSO) as an "environment within which humans may interact through simulation(s) and/or simulators at multiple networked sites using compliant architecture, modeling, protocols, standards, and databases." Within the synthetic environment is a synthetic natural environment (SNE) that represents the natural environment at a specific geographic location (geospecific) or at a typical class of geographic locations (geotypical). An SNE consists of the natural physical environment surrounding the simulation entities including land, oceans, atmosphere, near-space, and cultural information.

The collection of data is the most demanding and energy-consuming phase. Planners need to define categories of landscape data or classification of basic groups. Categories of landscape data could be labeled as data variables. A list of data variables determines the required data for calculation of models [MAR79].

The source data for the development of an SNE are acquired from a variety of repositories. The most common sources were developed by the Defense Mapping Agency (DMA); however, more recently, the National Imagery and Mapping Agency (NIMA) provides these data.

NIMA manages and provides imagery and geospatial information to national policy makers and military forces. It supports national security objectives by providing timely, relevant, and accurate imagery; imagery intelligence; and geospatial information.

A wide variety of data sources exists that cannot be exhausted in the chapter but can be found by examining the provided websites. The most common data sources for use in modeling and simulation include the digital terrain elevation data (DTED), interim terrain data (ITD), digital feature analysis data (DFAD), imagery data, and cartographic data. These key data sources are defined in the following section.

DTED is a digital elevation model of the terrain surface configuration characterized as a uniform matrix of elevation values for a specific geographic area. It provides basic quantitative data for all military systems that require terrain elevation, slope, and/or gross surface roughness information. DTED level 1 is based on a grid spacing of 100 meters; level 2 is 30 meter spacing, level 3 is 10 meter, and level 4 is 3 meter grid spacing. Smaller grid spacing provides a more accurate sampling of the specific geographic area but at the expense of larger data storage than required for larger grid spacing over the same area.

ITD is an interim product for digital terrain analysis that was provided until the objective tactical terrain data (TTD) source could be fielded in its final format. The ITD provides cultural/terrain feature models and is composed of the following six thematic layers: vegetation, surface materials, slope, surface drainage, obstacles, and transportation.

DFAD consists of selected natural and man-made, plain-metric features classified as point, line, or areal features as a function of their size and composition. Each feature is assigned an identification code and is further described in terms of composition, height, length, and orientation (e.g., lakes, rivers, forest, glaciers, urban areas, predominant towers and bridges, roads, railroads, and navigable waterways).

Weather and other environmental data can be obtained from systems such as the Master Environmental Library (MEL). The MEL is a wide area network-based data discovery and retrieval system that provides access to geographically distributed oceanographic, meteorological, terrain, and near-space information and data. The MEL will eventually host a models and algorithm catalog that not only addresses the atmosphere and space domains but also will be expanded to all environmental areas [BRA98].

Proper terrain modeling for combat simulation ensures that simulated combat forces execute tasks and missions in accordance with the landscape criteria. A synthetic environment should have an impact on fire, movements, and maneuver. With DRM, it is possible to project natural (forests, rivers, swamps, altitude, elevation, etc.) and man-made obstacles (urban areas, dams, roads, bridges, etc.), which have a direct impact on simulated entities and their activities. The interactions between forces and terrain are determined before start of computer-assisted exercise (CAX), so during the execution impact these interactions are a part of the automatic relations in the data base. The complete modeling of entities within a synthetic environment can be divided into the following components:

- Environmental parameters that represent, for example, the terrain surface, three-dimensional (3-D) model structures and states, and other SNE characteristics
- Environmental effects models that simulate the ambient SNE effects (e.g., atmospheric and transmission effects) and local SNE effects (e.g., smoke and rain)
- Component models that simulate entity component capabilities (e.g., sensors, movement platforms, and weapons)
- Behavior models that simulate the entity decision processes that result in observed entity actions that feed back to the environmental ground truth [BRA98]

Because of technical limitations of combat simulation models, it is necessary to prepare data bases of landscape and regional structure in the area of operations, as a part of the information system. The main components of the simulation are therefore a digitized terrain map and geographic information systems (GIS) with all the necessary environmental data. A digitized terrain map simulates visibility lines among units and visibility in general because of different terrain factors, and it provides a structure of terrain inclination [SAV00]. GIS enters in the simulation of all kinds of natural and synthetic objects and their effects on the operational units such as the following:

- Obstacles and various fortifications (antitank ditches, tranches, and mine fields)
- Natural obstacles (swamp, lakes, rivers, and canyons)
- Terrestrial vegetation (forests and bushes)
- Other objects (bridges, roads, and railway lines)

To support the unambiguous description of environmental data, the DMSO have been sponsored an ongoing project called The Synthetic Environment Data Representation and Interchange Specification (SEDRIS) to support the authoritative representation of the physical environment within the modeling and simulation world. SEDRIS is focused on solving the problems of achieving lossless and unambiguous environmental data representation and interchange across the heterogeneous distributed simulation community. The issue of environmental data representation and modeling is of great importance and wide interest regardless of how the simulation is constructed/operated (e.g., heterogeneous or homogeneous, distributed or not, multiprocess or single). It also persists throughout the simulation lifecycle from preexecution through initialization/execution to after-execution review and analysis [BIR98].

9.1.2 Data Coding Standard (DCS)

The SEDRIS data coding standard (SDCS) provides a mechanism to specify the environmental "things" that a particular data model constructs is intended to represent. That is, a "tree" could be represented alternatively as a point feature, an aggregate geometry, a data table, a model, or some combination of these. Which of these the data modeler (i.e., the data provider of the SEDRIS transmittal) chooses is orthogonal to the semantic of the "thing" that is represented—that is, what it is the representation of.

Having a standard way to accomplish something considered an infrastructure technology serves two very useful roles. First, it frees the user to concentrate on the more important application-level development, because there is no need to devote time and resources to designing the infrastructure. Second, it makes it possible for the users to communicate effectively and unambiguously through a standard mechanism. This, in turn, makes interoperability possible.

Another important role of standards and particularly international standards is to subject the technology to scrutiny by a wider, more diverse audience. This was the primary reason the SEDRIS organization elected to pursue the development of international standards. Finally, many in the community look for the "badge of approval" when it comes to embracing infrastructure technologies such as SEDRIS. For this reason, standardization through ISO/IEC was initiated.

Establishing formal standards was a key part of the SEDRIS development plan. Pursuing international standardization helped to ensure a broad base for applying SEDRIS technologies and opened interoperability opportunities in multiple national and international markets. However, developing formal specification standards was insufficient to realize all the interoperability potential that SEDRIS could provide. Establishing tested implementations, guidance and education documents, and data coding mapping documents was also required. Toward this end, in the spring of 2000 the Simulation Interoperability Standards Organization (SISO) established two product development groups (PDGs) to address technical implementation of the spatial reference model (SRM) and environmental data coding specification (EDCS) International Organization for Standardization (ISO)/International Electrochemical Commission (IEC) standards [SED07].

To support the unambiguous description of environmental data, SEDRIS specifies:

- A SEDRIS data representation model (SDRM)
- A spatial reference model (SRM)
- An environmental data coding specification (EDCS)

Although the SDRM addresses how to describe "environmental things" in terms of data modeling constructs that are meaningful to simulation developers (e.g., geometry, feature, image, topology, and data table), it explicitly avoids specifying "where" the "environmental things" are and enumerating all the "environmental things" that these data modeling constructs could be used to represent.

The SRM captures and unifies the spatial models used by SEDRIS, plus the many others that are not currently used but can be easily added. These models include inertial, quasi-inertial, geobased, and non-geobased (purely arbitrary Cartesian) systems. The SRM provides a unifying mechanism for specification and inclusion of any spatial reference frame and coordinate system. Its algorithms are designed to retain a high degree of accuracy during transformation and conversion operations (1-milimeter accuracy).

The EDCS provides a mechanism to specify the environmental "things" that a particular data model constructs is intended to represent. That is, a "tree" could be represented alternatively as a <Point Feature>, an <Aggregate Geometry>, a <Data Table>, a <Model>, or some combination of these or other SDRM constructs. Which of these the data modeler (i.e., the data provider of a SEDRIS transmittal) chooses is orthogonal to the semantic of the "thing" that is represented (and its location). The provision of such a "thing" in a SEDRIS transmittal must result in a shared understanding of "what the thing is and what it potentially means" to all applications consuming that transmittal. The purpose of the EDCS is to insure that environmental concepts align with the following requirements:

- Unambiguously defined
- Flexibly denoted and encoded and
- Easily bound in exchange formats and to programming languages

These environmental concepts are as follows:

- Classifications that define the type of environmental objects
- Attributes that define the state of environmental objects and
- Enumerants and units of measure that define how values of state are characterized

Classifications define types of environmental objects such as bridges, buildings, oceans, clouds, whales, trees and automobiles. Attributes define the state (sometimes called the properties) of environmental objects such as size, color, temperature, salinity, humidity, and frequency. Enumerants define a finite set of possible values of properties, such as {red, orange, and yellow, green, blue} for a color property. Units of measure define a nomenclature for characterizing specific quantitative values of properties, such as length, area, thermodynamic temperature, pressure, and electric potential. Denoting and encoding a concept requires a standard way of identifying the concept by use of a label or code [SED07].

The scope of the EDCS includes but is not limited to the following:

- Abstract concepts (absolute latitude accuracy and geodetic azimuth)
- Airborne particulates and aerosols (cloud, fog, and snow)
- Animals (civilian, fish, human, and whale)
- Atmosphere and atmospheric conditions (air temperature, precipitation rate, pressure altitude, wind speed, and wind direction)
- Bathymetric physiography (continental shelf, guyot, reef, and seamount)
- Electromagnetic and acoustic phenomena (acoustic noise, frequency, polarization, and surface reflectivity)
- Equipment (aircraft, artificial satellite, tent, train, and vessel)
- Extraterrestrial phenomena (comet, planet, and spacecraft)
- Hydrology (lake, rapids, river, and swamp)
- Ice (ice field, ice peak, ice shelf, and glacier)
- Man-made structures and their interiors (bridge, building, hallway, road, room, and town)
- Ocean and littoral surface phenomena (current, surf, tide, and wave)
- Ocean floor (coral, rock, and sand)
- Oceanographic conditions (luminescence, salinity, specific gravity, and water current speed)
- Physiography (cliff, gorge, mountain, and valley region)
- Space (charged particle species, ionospheric scintillation, magnetic field, and particle density)
- Surface materials (concrete, metal, paint, and soil)
- Vegetation (crop land, forest, grass land, kelp bed, and tree)

The EDCS supports the encoding and communication of qualitative and quantitative information associated with physical environments, both real and virtual. This task is accomplished by specifying a collection of nine EDCS dictionaries of environmental concepts. An EDCS dictionary is a list of EDCS dictionary entries, each of which specifies a single concept. Each EDCS dictionary contains entries of a similar nature; however, each entry is unique. Each EDCS dictionary entry consists of the following fields: definition, label, code, references and reference type, supplemental references, and other

EDCS dictionary-dependent information, such as value types, unit symbols, or unit equivalence classes.

The nine EDCS Dictionaries are as follows:

- EDCS Classification (EC) dictionary
- EDCS Attribute (EA) dictionary
- EDCS Attribute Value Characteristic (EV) dictionary
- EDCS Attribute Enumerant (EE) dictionary
- EDCS Unit (EU) dictionary
- EDCS Unit Scale (ES) dictionary
- EDCS Unit Equivalence Class (EQ) dictionary
- EDCS Organizational Schema (EO) dictionary and
- EDCS Group (EG) dictionary

Additional entries may be added to most EDCS dictionaries through the process of registration. The EDCS application program interface (API) supports conversion between real values given in different units of measure [SED07].

By providing a rigorous, extensible, internally consistent, and increasingly complete data dictionary from which to select environmental objects and attributes for use in simulation, the SDCS offers SISO an opportunity to "stand on the shoulders of giants" with regard to environmental data dictionaries. Additionally, a new version of the SDCS includes not only support for the full range of land, sea, air, and space environmental domains, but also support for localized fixed and mobile "things" of both military and civilian nature [BIR98].

During the decision-making process about the long-term usage of space, spatial researchers provide deliberate solutions for spatial planning. Because of that, they are an inevitable part of the decision process and enter a sensitive area of societal events on a cross road of societal interests and their needs, as well as economic lawfulness and policy. Researchers also legally define mechanisms to protect the natural environment. Decision-making theory is a relatively young science, which takes knowledge and ideas from many other sciences using operational research, probability theory, artificial intelligence, recognition of samples, economy, psychology, and theory of soft groups. The intense development started at the same time with developments of computer technology in 1970s. The first 10 years were dedicated to the theoretical research and creation of models and methods. In 1980s and 1990s, the decision-making theory has been divided to different functional areas, which are supporting decisions. Recently, focus was on group decision making, multicriteria decisions, multi-objects decisions, and soft decision-making processes [SAV00].

The contemporary world is characterized by a collection of data, information, and experiences, where fast and relevant analysis is the key for effective planning and decision making, which includes the management of economical, governmental, military, and ecological systems. Often data are incomplete, information inaccurate, experience insufficient, and the decision-making system demanding and complex. Human decision making is in these conditions extremely difficult and responsible. In this cases, computer-assisted systems offer support for decision-making processes [SAV98].

9.2 ATTRITION AND MOVEMENT

Perla [PER90] has emphasized that the war game is a model or combat simulation, which does not involve real military units, and where the course of events depends on decisions of players, between two opposite sides. The key words in this definition are players and decisions. War gaming is an experiment of interaction among people. Without players, it is a model and not a game. Focus on the human decision-making process demands a qualitative structured game, which will ensure conditions for shaping decisions and understanding their effects. Because of that, there is a need for formulation of objectives, scenario, data bases, rules, players, and analytical elements [PER90].

CAX is a training method where, with the help of simulation model as a training tool, preparation, execution, and analysis of commanders' decision-making process has been exercised. Regardless of type, a model can be deterministic or stochastic. The initial conditions of interactions among terrain data base and weaponry systems data base are determined. At the beginning of the exercise until the end, the course of events is uncontrolled, pending the decision-making process and communications between command group and subordinate units on work stations.

The purpose of the decision-making process is to improve the future state of an individual or organization; in the area of security planning, the aim is to improve the future state of society. The decision maker is an individual (or group) that is not satisfied with the current state or the future perspective of that state, and at the same time he/she has willingness and authority to modify the current state [GAS96]. The decision maker makes decisions on the basis of an interpretation of the results from the simulation (Figure 9.2). If the decision maker is not satisfied with the results or if he or she cannot make a decision, than the whole problem needs to be defined again, in respect to decision maker's remarks. The decision maker has to regulate undesirable disturbances of the system to achieve higher personal and group satisfaction [BEN78].

An operation is the science of employing military assets. It is focused on maneuver and firepower and takes advantage of available forces to inflict maximum damage to the enemy for minimum damage in return. Aggregate simulations are used to determine the outcomes of operational decisions, in a way that is plausible and statistically reasonable. In this application, the simulation must account for the movement, sensing, and exchange of weaponry; damage effects; and communications.

Synthetic entities (SEs) are artificially created active simulation components that represent entity and human behavioral characteristics in simulation both directly and indirectly. SEs may be main components of a simulation that

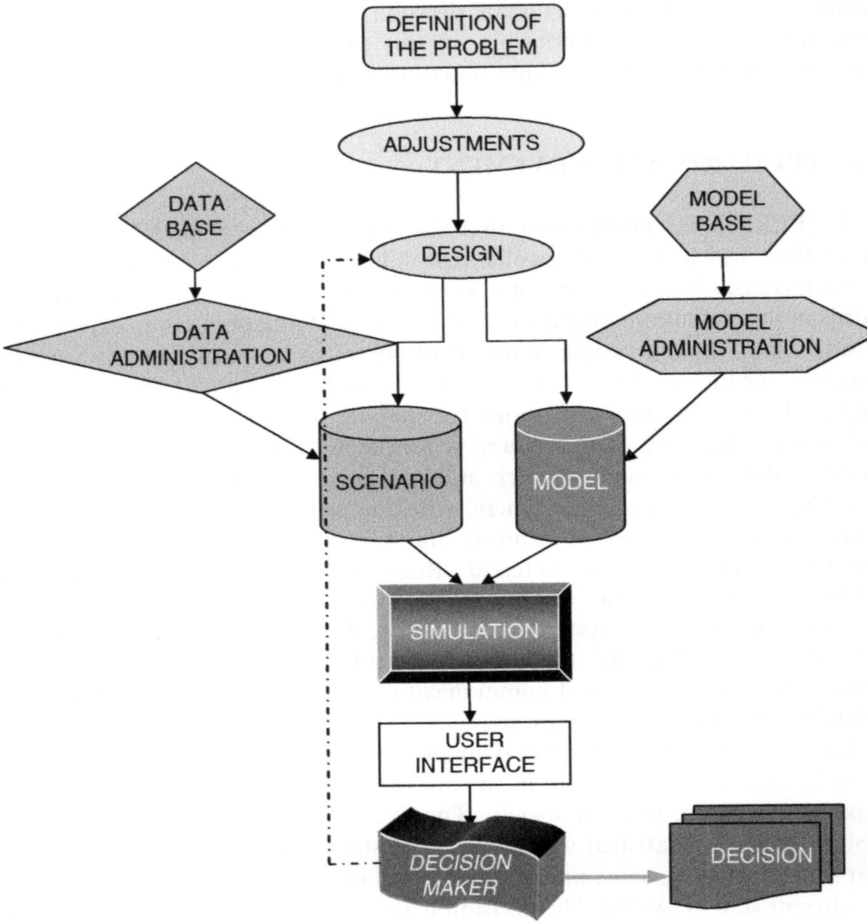

Figure 9.2: Creation of training conditions for decision-making process.

directly produce stimulus for users and other SEs while actively interacting with other simulation components such as the SNE. SEs are both complex and demanding while providing a significant hurdle to the modeling and simulation community for their development and implementation. Some example representations include individual soldiers, tanks, fighters, platoons, companies, battalions, squadrons, wings, and command and control posts, and so on.

Computer-generated forces (CGFs) are the most popular instantiation of SE. CGFs model individual combatants and combat organizations while executing tactical behaviors or making tactical decisions in simulations.

In many cases, these behaviors and decisions are made, in part, by human operators. If so, they are typically called semi-automated forces (SAF). SEs executions that fully automate decision makings through algorithms or artificial intelligence (AI) technologies participate in a simulation as automated forces (AF). Additionally, SEs may represent the existence of other human participants (e.g., crew members) in a simulation space, which allows the actual human players in a simulation exercise to interact with other components such as synthetic environment or other SEs.

Some of the earliest deterministic attrition models described continuous fire combat, where one side erodes the combat power of another, at some fixed rate over time. The most prevalent example of a deterministic combat model are the Lanchester equations, which were first published by a Victorian era engineer who developed a mathematical force-on-force theory of combat in 1914. Lanchester theorized that each side in a combat duel degrades the other side at some rate proportional to its own remaining size multiplied by the firing rate of its shooters. Using differential equations, Lanchester equations prescribe such results as the ultimate winner of a contest between combatants, the time required for a duel to conclude, or the size of each force remaining at a duel's conclusion. This model is the basis for most of the current attrition-based combat simulations in use today. Traditional stochastic combat models represent combat as a chain of independent events (each with their own probability of occurrence) or as sets of basic interaction equations, in which random variables represent operational processes [CAR04].

Combat as a challenge was always in the focus of people. Regardless of their position during many engagements, they achieved smaller or bigger success.

9.2.1 Lanchester Equations

Combat modeling means the formation of a real combat model or simulation of real combat, as well as its solution with the mathematical and statistical methods. The first steps were done by Lanchester and Osipov in 1914, where both of them used similar differential equations for combat descriptions. Both of them were focused on mass and concentration of forces. Opposition forces have been displayed by quantitative value, which could be observed as a variation during combat. That variation is defined as attrition. The speed of attrition is determined by attrition coefficients, which could be defined for each engagement separately. The combat model describes real combat, and it is important to assess attrition coefficients with the help of probability theory and statistics. A rapid development of computer technology at the end of the 20th century allows new possibilities for the future application of complex Lanchester equations.

Since it was first proposed during World War I, the Lanchester Combat Model in various forms has been used to address quantitatively the competition between opposing forces. The model not only provides insight as to who the victor may be, but also it can approximate how much of each force remains, how long the competing forces remain engaged, and how the changing conditions of battle can impact the outcome. Consider two opposing forces—red force and blue force—with respective numbers of troops $R(t)$ and $B(t)$ at time t. Although time is often measured in hours or days, any time frame would suffice. Over time, the numbers of troops available for combat change for various reasons—perhaps the most important is combat losses.

Combat losses are indicative of the ability of one force to eradicate the other. When conventional forces meet, combat losses are considered to occur at a rate proportional to the number of troops of the opposing force. Assuming that the blue force and the red force can inflict combat losses on each other at proportional rates of k_1 and k_3, respectively, we can model the changes in their troop numbers by

$$\frac{dR}{dt} = -k_1 B(t), \quad \frac{dB}{dt} = -k_3 R(t)$$

Whereas combat losses account for much of the change on the battlefield, noncombat losses can degrade troop availability. We assume that the forces sustain these losses at rates proportional to their own numbers. Thus, if the proportionality constants are k_2 for the red force and k_4 for the blue force, the model becomes

$$\frac{dR}{dt} = -k_1 B(t) - k_2 R(t), \quad \frac{dB}{dt} = -k_3 R(t) - k_4 B(t)$$

The final component of this model is troop reinforcement. With the red force and the blue force getting reinforcements at rates of $r(t)$ and $b(t)$, respectively, the combat model with initial conditions becomes

$$\frac{dR}{dt} = -k_1 B(t) - k_2 R(t) + r(t), \quad R(0) = R_0$$

$$\frac{dB}{dt} = -k_3 R(t) - k_4 B(t) + b(t), \quad B(0) = B_0$$

Perry [PER06] has realized that collective combat between a red side of strength R and a blue side of strength B is described by the following equations:

$$\frac{dR}{dt} = -k_B B(t), \quad R(0) = R_0$$

$$\frac{dB}{dt} = -k_R R(t), \quad B(0) = B_0$$

which result in the equation of state:

$$\frac{(R_0^2 - R^2)}{(B_0^2 - B^2)} = \frac{k_B}{k_R}$$

The quadratic form results in the system of equations known as the Lanchester Square Law.

Individual combat on the other hand is described by the following equations:

$$\frac{dR}{dt} = -k_{BR}B(t)R(t), \quad R(0) = R_0$$

$$\frac{dB}{dt} = -k_{RB}R(t)B(t), \quad B(0) = B_0$$

With its equation of state:

$$\frac{(R_0 - R)}{(B_0 - B)} = \frac{k_{BR}}{k_{RB}}$$

The linear form of the above equation results in the system of equations known as the Lanchester Linear Law. Over the intervening years, there have been many attempts to validate the use of Lanchester's equations to describe combat outcomes through analysis of historical data. These efforts have largely been unsuccessful. Part of the reason for this lack of success is the failure of subsequent users of Lanchester's theory to understand its inherent assumptions, constraints, and limitations. Typically, most users assume that the fighting will be entirely collective and that both sides will attrite each other evenly, with strengths asymptotically approaching zero. Yet random chance is a factor in warfare whose effect can be modeled by treating attrition as a stochastic process.

Although real battles are nonuniform in space and time, neither parameter appears in Lanchester's equations. This results from an implicit assumption that all the forces of each side can engage the other (or in other words, the ability to engage is not target limited) in the derivation of the square law.

However, from a study of the evolution of the Battle of Trafalgar, Lanchester realized that a battle is more properly viewed as a series of concurrent and consecutive sub-battles separated by space and time. Lanchester proposed that each of these sub-battles be described using his square law and that the overall casualties of the battle are obtained by summation of those losses. A recent review of this work has shown that each of those sub-battles can also be deconstructed into a set of smaller sub-battles, and the outcome of the battle can be shown to be strongly dependent on the ability of commanders to fragment the battle to their advantage.

The successful application of Lanchester's equations to describe actual battle results first requires an understanding of the structure of that battle to enable a hierarchical decomposition of the battle into smaller sub-battles until the assumptions underpinning Lanchester's equations are met by those sub-battles. It could be said that this decomposition results from the application of command and control on the battle's evolution. At any rate, the outcomes of those sub-battles can then be determined and the results propagated into the starting conditions for the following group of sub-battles. This process is repeated until the end of the battle. This is the process that Lanchester envisaged, but it is not how most practitioners use his theory [PER06].

9.2.2 Stochastic Processing

Stochastic modeling is based on conditional probability that certain social event or process will exist or appear. Starting in the 1970s, the widespread use of digital simulation facilitated the inclusion of randomness and uncertainty in analytical models. A model that explicitly includes randomness and uncertainty is called a stochastic model; stochastic models are usually created by modifying one or more of the terms in a deterministic equation with random draws from some probability distribution. These models are run a certain number of times so that some effects of randomness or uncertainty cancel out and model results converge to stable, aggregate statistical values [PER06].

Thus, stochastic simulation has to be executed many times, because each simulation has many possible outcomes. This type of simulation has been used in demographical and ecological simulations, where many components are modeled by statistical criteria. Stochastic simulations are particularly useful for statistical analysis, because many systems' behavior can be explained only by statistical notion. This type of modeling is fundamental for combat modeling and the analysis of the decision-making process during CAX, which enables desirable observations of combat simulation with the statistical and probability models.

Stochastic models have three additional significant assumptions. The first assumption stems from a basic problem in modeling uncertainty: Because uncertainty implies an incomplete knowledge of the input data required for a model, then some input parameters must be random variables. The second assumption flows from the first: Because the data itself are to some extent uncertain, so are the interactions between inputs. The parameters must therefore be treated as independent random variables. In other words, complex chains of causality in the operational processes being modeled are considered inconsequential. It is assumed that most processes can be modeled as either independent events or as chains of simple causality. Third, modelers frequently assume that the distribution of outcomes for these independent random variables is not so skewed that a relatively small number of model runs will mitigate variation in the random variables and produce statistical convergence.[2]

Traditional stochastic combat models represent combat as a chain of independent events (each with their own probability of occurrence) or as sets of basic interaction equations (with random variables representing operational processes). Combat models generate events on the basis of algorithms in data bases. Independent events are usually the result of interactions between characteristics of digital terrain as well as attributes of combat systems on

[2] In other words, the Central Limit Theorem applies. See http://mathworld.wolfram.com/CentralLimitTheorem.html, accessed 29 Feb 2008.

the red and blue side. Different simulated engagements support the decision-making process, where the final outcome of the battle depends on timely reactions and the quality of decisions of a training audience.

9.3 CHALLENGES IN THE QUANTIFICATION FOR NONKINETIC WARFARE

Existing simulation models are designed primarily for combat simulations. The status of combat models depends on the interaction between weaponry systems or kinetic warfare and the impact of space on these systems. Most combat models have no data about civil structures, refugees, or activities of nongovernmental organization (NGOs). Because of that, during combat simulations we can observe the following variables: consumption of fuel, ammunition, water, number of causalities, injured soldiers, and so on. With these observations, is not possible to assess the efficiency of the crisis response operation (CRO). The characteristics of contemporary CRO, where almost 70% of all activities are nonkinetic, are forcing us to improve our knowledge, data bases, and the way of operation we analyze the. Instead, to measure the effects of one's own forces on opposite forces and their structure, we have to measure the impact of peace forces on the affected society. It was already mentioned that societal security could be observed through societal security dimensions and their potentials. Those potentials are the result of the local capability on a municipality level, to predict and manage possible threats to the security dimensions and to provide security for the local population.

Obviously, gaps exist in modern simulation models, which need to be filled. There is a need to create additional data bases in the form of information systems (IS) to fill the data gaps about the area of complex emergency. During a preparation of CAX, technicians prepare the synthetic environment like in combat simulations, and at same time they prepare data in IS, which has to be more focused on society as a factor (detailed description of the area of complex emergency, concise analysis of terrain, national, economical and historical root causes of the conflict; multimedia contents about key phases of the conflict). CRO is connected to the United Nations (UN), thus, structural parts of IS are also basic political documents, which provide fundamentals for execution of operation to the CAX participants. Political documents consist of scenario, road to crises explanation, cease fire agreements, resolutions of UN Security Council (SC), standard operating procedures (SOPs), rules of engagement (ROEs), different agreements for intervention of peace forces, law of war, and other documents that can help to simulate peace operation. All these documents should be available to the CAX participants on the exercise intranet pages to ensure the correct execution of decision-making process and exercise in its entirety. To activate exercise processes, exercise control uses a preplanned list of simulation event called the master scenario event list (MSEL), which is linked to the training objectives. Each event, incident, and inject has predicted participants or training audience (e.g., paramilitary units, civil police, military units, and civil organization) and expected outcome as a result of decision-making process and interactions among the training audience (TA). The nonlinear dynamics of recent operations are a primary source for modeling the situational and synthetic events in the simulation model. With creation of directive, informative, and interactive incidents[3] for peace forces in connection with the synthetic environment, it is possible to coordinate both kinetic and nonkinetic activities (Figure 9.3) to manage a complex situation until the desirable end state of operation is achieved.

Modern society is a sociotechnical system with the necessary social functions associated in different combinations. A consequence of that is a rapid dissemination of effects of a certain event and the uncontrolled developments of a situation among linked security dimensions. Society has become more vulnerable, confronted with different and unpredictable threats. That is visible on local level, where security threats are real, and local authorities need to deal with them on a daily basis [NIL02].

For the proper quantification of nonkinetic warfare and capability assessment of local authorities, in a synthetic environment, it is necessary to determine the functions and societal areas for analysis before starting a CAX. They could be connected with the object of risks, areas of risks, and how the local authorities are dealing with possible threats and developing their own skills for decreasing vulnerabilities. The vulnerability of the local community correlates with its capabilities to manage complex security conditions and therefore also a level of survivability for the population. Local authorities with the acceptable level of vulnerability could be defined as capable and flexible enough. The most appropriate statistical methods for analyses is the multivariate analysis of local capability for societal reconstruction and the cluster analysis that allows the researcher to define typology and clusters of municipalities by groups of variables.

For an appropriate initial situation in the area of complex emergency on CAX, it is possible to use potential of societal security dimensions of the affected society. The methodology of quantitative secondary analysis of available societal variables, which are collected remotely in the area of interest, represents the empirical results of the structure, events, and processes in the area of complex emergency.

Example 9.1

At the end of April 2008, 30 municipalities of Kosovo were investigated as a case study of local capabilities for societal reconstruction. The variables for

[3] The directive incident triggers certain activity in the TA, for example, an interview with the media about developments of the operation or meeting with the NGO and civil military cooperation CIMIC representative. An informative incident provides information to the TA, for example data about internal displaced persons or weather forecast. An interactive incident ensures situational awareness, list of decisions, or different methods for observing complex situations.

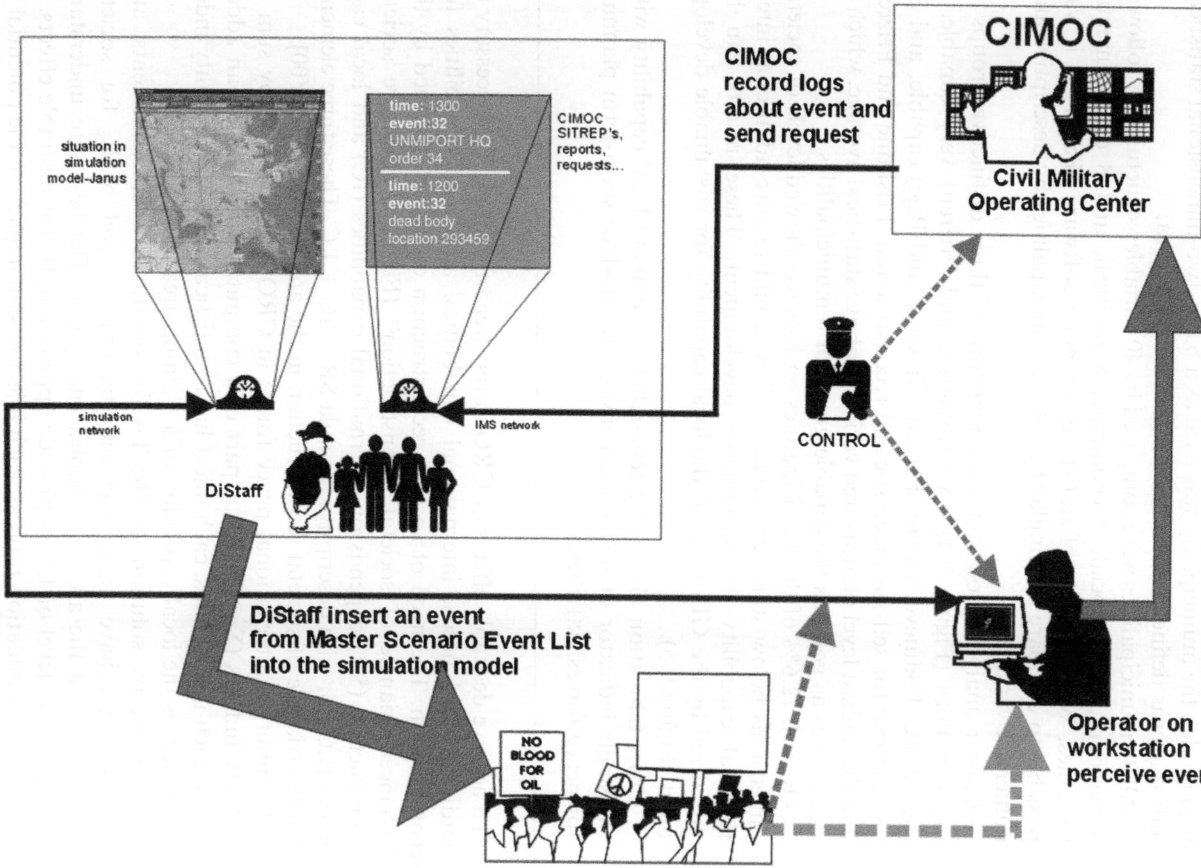

Figure 9.3: Injection and execution of task during a peace operation.

secondary analysis were collected from open sources on the Internet, from the Statistical Office of Kosovo, and from OSCE reports, HCIC, and some other humanitarian organizations. For populating the multivariate analytical model, the municipality was used as a basic, two-dimensional statistical unit that was defined by the name and size of the area.

All 30 municipalities were compared by 40 variables structured as follows: 8 demographic, 8 social, 8 economic, 8 political, and 8 environmental variables. With the help of statistical computer software SPSS 11.5, it was possible to compare the similarities of the municipalities by descriptive and numerical variables.

Cluster analysis produced a clear picture of three distinctive groups of municipalities (Figure 9.4): two (Prishtine and Prizren) very capable, 6 (Mitrovice, Podujeve, Gjilan, Ferizaj, Gjakove, and Peje) capable, and 22 less capable for self-sustained societal reconstruction. Prishtine and Prizren had the highest level of measured variables above statistical average, which is three times as good as the capable group of 6 municipalities.

Overall, 72.6% of the municipalities on Kosovo showed levels of societal dimensions below statistical average. They would require additional international capability for sustainable development. These results can be visualized by the GIS layer of the local capability for sustainable development (Figure 9.5).

Cross tabulation allows the researcher to compare local capability with the single indicator of societal dimensions to produce short-term planning for peace forces activities.

To achieve desirable effects of CRO in the area of interest, it is necessary to coordinate local needs, international capabilities, and local capabilities for reconstruction. The need of societal reconstruction could be explained by the "Enhanced triangle of societal reconstruction" (Figure 9.6), where societal reconstruction (SR) depends on international capabilities (IC) and local capabilities (LC), or with descriptive formula SR = IC + LC. These three elements form the triangle of societal reconstruction in the area of CRO [DOY00].

The input of international peace forces in CRO is represented by sum of capability indexes (Σci), which is part of a computer simulation as an added value of each inject. Thus, each inject has a statistical value or capability index that influences the local capability of a training area on CAX.

The qualitative solutions of the TA for situational events, incidents, and inject on CAX have an impact on quantitative level of IC for societal reconstruction in the area of complex emergency. The key is to understand the quantitative level of LC to provide appropriate IC for desirable effects.

During the simulation process, each inject has an impact on the potential of security dimensions (demography, social situation, economy, political situation, and environment). To quantify nonkinetic activity, it was necessary to create a capability index for each inject, with the help of weights and elements

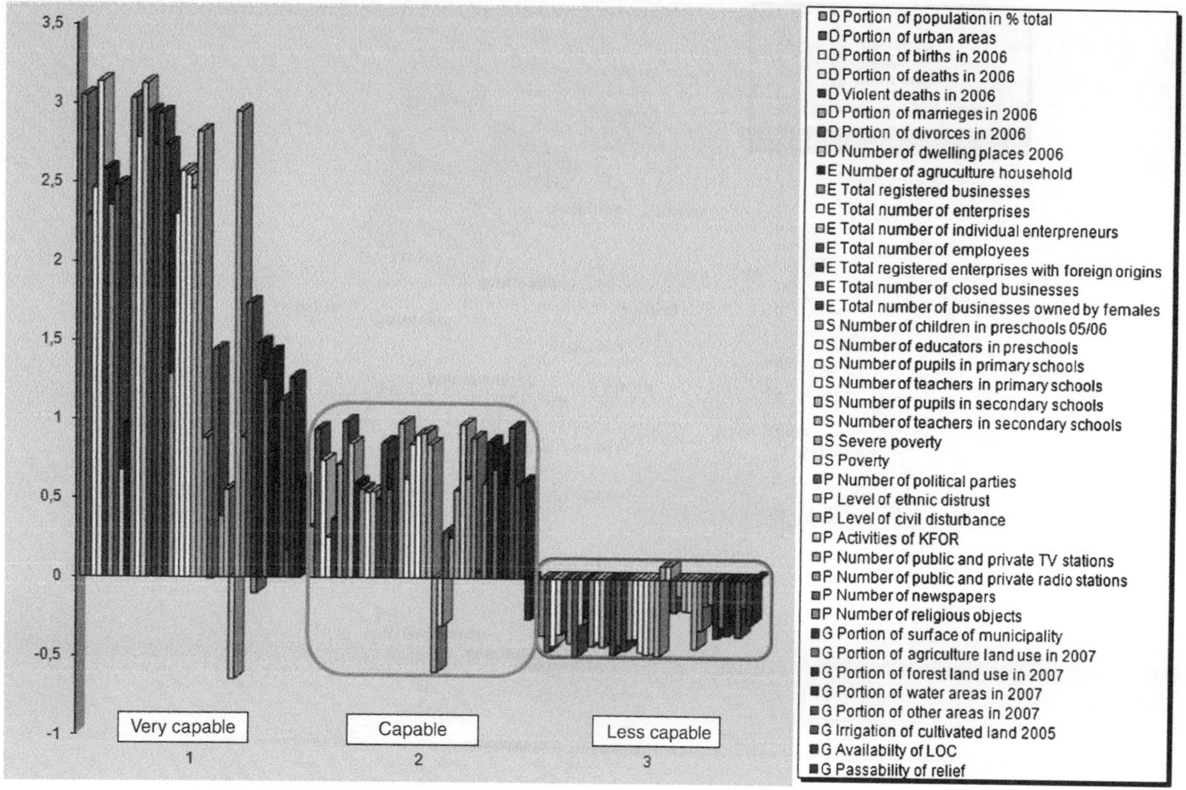

Figure 9.4: Local capability for sustainable development of Kosovo in 2008.

Figure 9.5: GIS layer of local capability for the sustainable development of Kosovo in 2008.

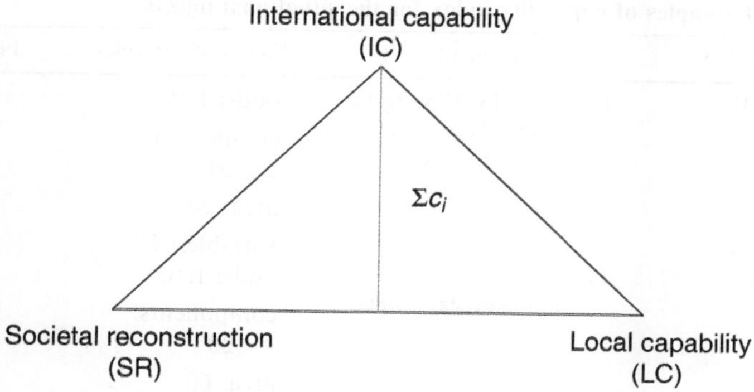

Figure 9.6: Enhanced triangle of societal reconstruction.

of influence in form of statistical value. The decision-making process of TA produces a certain outcome. The comparison between expected and real outcomes creates conditions for adding a capability index to the initial level of local capability. The total sum of changes on the security dimension's potential shows the level of LC in the area of operation; however, it also evaluates the efficiency of peace forces components and the effectiveness of the operation in a certain time and space.

So the key question is what elements have the capability index in its structure? With the help of log-odds scale, it is possible to define the statistical value for each inject. Besides the log-odds scale (Table 9.1), we have to take in to account the following weights of influence (value from 0 to 100):

- Number of involved components in the inject management process (military 20, police 20, civil organization 20, local authorities 20, an international community 20),
- Size of the area on which inject has influence (pinpoint: 20, village up to 300 people: 40, city: 60, partly area of responsibility: 80, and whole area of responsibility: 100),

Table 9.1: Log-odds scale and its numeric nomenclature.

Metric name	Scale	Odds	Probability (approx)
1 chance	0	1:1	0.5
	−1	1:10	0.1
	−2	$1{:}10^2$	0.01
1 milli chance	−3	$1{:}10^3$	0.001
	−4	$1{:}10^4$	0.0001
	−5	$1{:}10^5$	0.00001
1 micro chance	−6	$1{:}10^6$	0.000001
	−7	$1{:}10^7$	0.0000001

Table 9.2: Examples of capability index for the situational injects.

Time of Injection	Incident	Values of variables	Final value
0601000 NOV	LG 3 LOGISTICS- Bridge damage; Accident of POL road tanker.	– odds: $1{:}10^3$ – components: 2×20 – area: 20 – variables: 50	**0,11**
061400 NOV	Larger demonstration More than 300 people.	– odds: $1{:}10^3$ – components: 3×20 – area: 60 – variables: 75	**0,195**

- On how many security dimensions the decision making process has an influence (political 20, economical 20, social 20, demography 20, landscape 20).

The total sum of all mentioned variables is used to calculate the capability index for each inject in simulation (Table 9.2). The obtained value represents the added value to the potential of local capability for societal reconstruction of affected society.

With the proper number of injects in the CAX, it is possible to measure the efficiency of the TA, which executes decision-making process on a simulated CRO. The multivariate analytical model allows an analyst to observe and compare the performance of the TA at a certain time and space (Figure 9.7).

The final quantitative analysis of the performance of the TA is in the form of a percentage of reached local capability compared with the possible level of local capability in one group of municipalities (Figure 9.8).

9.4 AUTOMATED FORCES

As was already mentioned, automated decisions made through algorithms or AI technologies participate in a simulation as AFs. They are part of a system that ensures combat simulation with the ability of quick changes of the parameters and scenario. The input in the system is a digital terrain model, force disposition, equipment, and operation's plan. The result is the entire engagement process, display of units' dispositions, consumption of material, damage, and losses. The average simulation model is object oriented and event driven. The model uses the following objects: combat elements, command elements, interactive elements, and environmental elements. The system functions in accordance with certain principles [KNO98] as follows: "Each object is in certain point in time in certain state. Object can receive or send information. A change of state is possible only if something happened. If the object has received new information, than he can

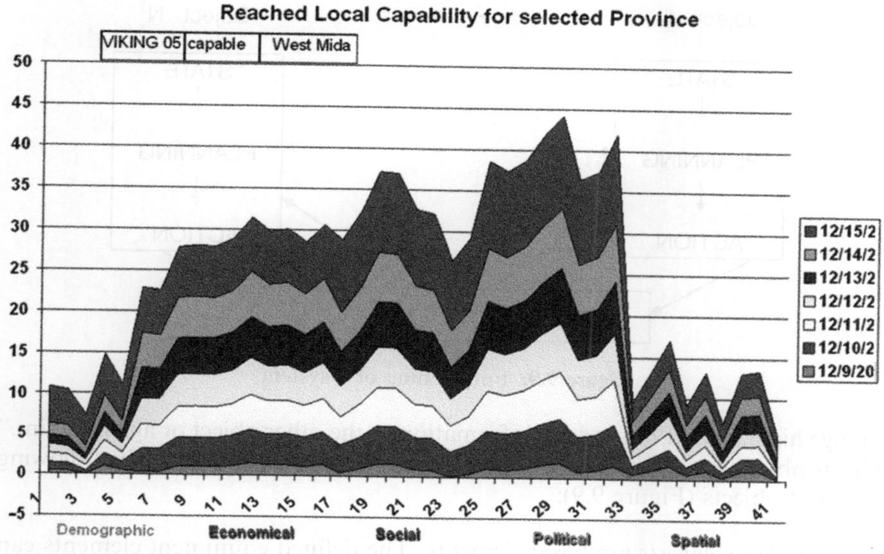

Figure 9.7: Development of local capability on peace support operation (PSO) CAX Viking 2005.

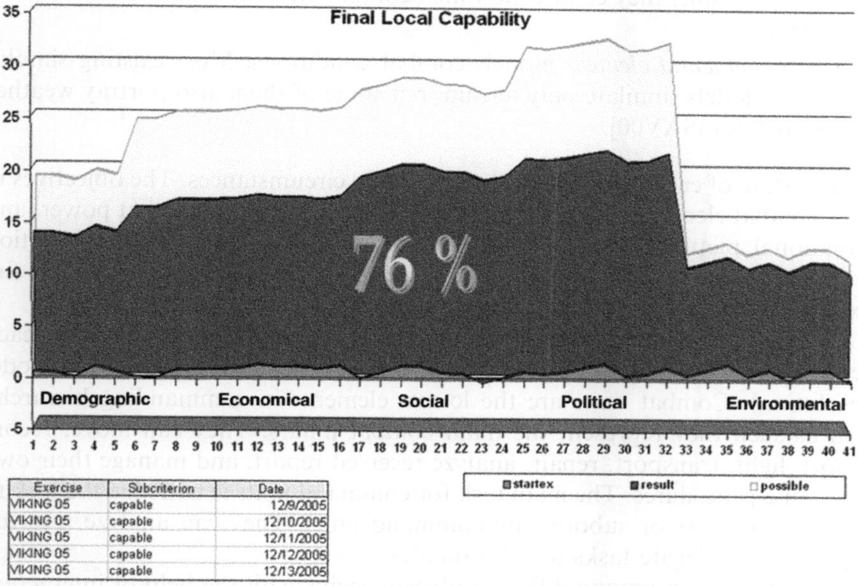

Figure 9.8: Final comparison of local capability on PSO CAX Viking 2005.

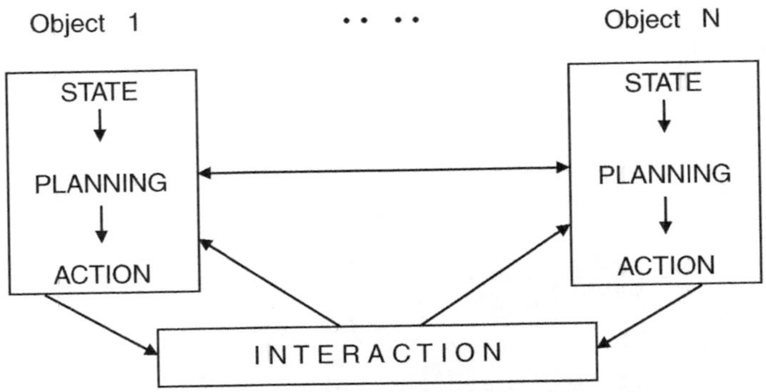

Figure 9.9: Functioning of a system.

change his state and/or send the information to the other object or again to him." This is obviously characteristic of AF, which has been classified in the following group of objects (Figure 9.9):

- *Combat elements* are basic elements. The defined equipment elements can move, shoot, recognize opposites, receive and send report, load and unload equipment, build obstacles, repair, and so on.
- *Command elements* generate orders for subordinate units. They can receive and send reports.
- *Interactive elements* manage natural interactions among combat elements. For example, they can define which combat element is visible to the other elements.
- *Environmental element* models combat conditions. Most existing simulation models simulate only terrain, but some of them also portray weather dynamics [SAV00].

The state of each object is determined by its circumstances. The objectives in the "combat elements" group could be defined by location, combat power, and situational picture. Based on state of certain element, plans for the action should consist of, for example, movements or engagements with the fire exchange. The plan could be created by management module or by user.

A group of combat elements consists of different combat units for each service, which could be divided to the single vehicle (depends on model resolution). Combat units are the lowest elements in commanding hierarchy and as such they represent the main combat pillars. They can move, recon, report, fight, transport, repair, analyze received report, and manage their own operating procedures. The main task for command units is to define the task to the combat units or subordinate command units. They can analyze received reports and delegate tasks to subordinates.

The interaction among different objects begins with the help of interaction modules, which are active when a certain object sends to the other one some information about event (for example, "fire at you"). The interaction module counts the results of each interaction, which could be the results of reconnaissance, effect of the weapon on target, and content of received report. These results depend on the state and action of an involved object (combat units or command units) and also on the data about an object's environment.

The Object environment creates conditions for other objects. It could be defined by factors of terrain such as covering, heights, and road network. It is possible to extend the environment with the conditions of light (day or night) and visibility. Modeling of the framework conditions as an object provides unity of dynamic modeling. As an example, a combat unit can change, with its actions, the state of another object by firing or laying a mine field [SAV00].

Command automata execute the decision-making process in the simulation model. They steer the behavior of combat elements because of the following:

- Tasks
- Weapons (combat capability) and combat power of units
- Current condition of a unit

These automatic elements of command and control provide orders to the units for the requested combat activity. An automatic element activates the appropriate orders to subordinate units, depending on the situation. The module calculates the best moment for the activity in the area of responsibility of an individual unit. The active automatic elements of command and control monitor the activities of units and their results of tactical tasks execution (attack, defense, movement, etc.).

The application of commanding automata has the following benefits:

- User defines only the objective of a certain task. The automated system creates several steps for execution.
- User delegates task to appropriate commanding unit and not to a singular combat element.
- Altogether, the lower number of inputs is necessary to accelerate the execution of combat simulation.

The weaknesses of commanding automata include the lack of transparency and acceptability, in particular when the decisions from AF do not satisfy the expectations of users. The users can communicate with the system to control the AF.

The simulation system works on the basis of models, which monitor the activity of singular elements in combat. The foundation is the terrain data base that consists of all information on the terrain and with the consideration on visibility rules, such as optical visibility, range and resolution of radars, characteristics of optical sensors, and visibility angle. For the movements, certain rules should be considered as follows: on roads, on terrain, for tactical movements, on swamp and trough the forest, river crossing and breaching the obstacles, flying on different heights, or sailing on surface or subsurface. The data base should define the usage of weaponry systems, vehicles, infantry entities, air forces, naval forces, and direct and indirect fires, which could be managed by the special editor. That editor defines decision making as well as the organizational structure and events that follow certain orders [SAV00].

AFs controlled by artificial intelligence are becoming more common for all categories of military constructive simulations. They can introduce considerable cost reduction because the more entities that are represented as AFs, the fewer simulation operators that are needed. They may also be very useful to control situational and especially opposing forces in a CAX. However, there is still a debate about AFs. The more commonly AFs are used in constructive simulations, the less predictable the results from simulations and therefore the flow of an exercise become [i.e., main events and incidents list (MEL/MIL)]. That is perceived as a risk by many exercise planners because MEL/MIL is the main tool to ensure that the exercise objectives are achieved. We believe that this is not correct for two reasons: 1) AFs are generally implemented as a capability that exercise controllers can have it switched on/off for individual entities/units and 2) dynamic scripting can be used to create new incidents and injections anytime based on the outcomes of a simulation.

Another approach for more controlled simulations involving AFs is to introduce SAFs instead of AFs. The autonomy of SAFs is limited, and they require human reasoning for critical decisions. Therefore, they are more under the operator control, and exercise controllers can still steer an exercise although automated entities exist in the simulation.

AFs require realistic behavior models used for computing the possible outcomes of the decision-making process for humans, crowds, missions, units, and so on. Behavior models use both AI techniques and domain knowledge, such as tactics and doctrine. Various AI-based decision-making techniques are available, such as decision trees, which assess the situation, and make decisions like the following:

- A computer-generated troop decides to shoot during a covert operation.
- A computer-generated crowd decides to disperse.
- A computer generated combat air patrol (CAP) mission decides to scramble.

One solution for AFs is to represent each AF entity as a software agent that can be characterized by the following features:

- Situatedness: The interaction of entity with its environment and its communications with the other agents.
- Autonomy: The ability to decide and control its own actions and capability.
- Flexibility and adaptability: The ability to adapt to the developing and/or unexpected situations.

Decision making and planning in AI is regarded as a search problem for the best course of action (CoA) among all alternatives. The goal is to find the best sequence of actions, which can take the system from its current state to the desired end state. This approach is called also task model or belief-desire-intention model. Of course, the major challenge in this approach is the number of alternatives. In some cases, the number of alternatives is too many to be searched for the best in a reasonable time even for a single entity and by using the fastest computers. Various techniques are available, which cannot guarantee the best solution but a feasible one within a reasonable time period. AI decision making is typically based on many things such as follows:

- Plans
- Missions
- Capabilities of the AFs
- The behaviors and status of the other entities in the simulation
- The other situational parameters, weather, terrain, social and economical conditions, etc.
- Domain-specific knowledge

Reasoning and decision-making systems can be in different levels, such as follows:

- Strategic
- Operational
- Tactical
- Entity level

In addition to the variety in the level of reasoning, AFs can represent various types of entities, such as follows:

- Individuals (soldiers, civilians, etc)
- Crowds
- Missions
- Units
- Organizations (a transportation/energy network, UN, nongovernmental organizations, etc.)
- Facilities (hospitals and fire brigades)

AI decision making can also be categorized as deterministic or stochastic. For example, a CAP mission scrambles the closest enemy air mission if it is within a certain radius. Some prioritization, such as, "first scramble enemy tankers," can also be given. This is a deterministic approach. The outcome of this process is the same for every trial for the same situation. In a stochastic case, random numbers may be used. For example, when there are multiple enemy air missions detected, an automated CAP mission selects randomly one of the enemy missions within a given radius to engage with. The outcome of this selection may be different at each trial for the same situation.

All these factors (i.e., the level of reasoning, the type of the entity represented as an AF, stochastic or deterministic decision making, and fully automated or semi automated forces) determine the complexity and practicality of the autonomy solutions for the computer-simulated entities. Note that AI-based decision making (i.e., searching for the CoA), is not the only issue for AFs. The crucial aspects for AFs are reasoning systems, knowledge acquisition, and representation and behavior modeling. Accurate, validated, adaptive, and scalable behavior models, as well as the collection, maintenance, and querying of accurate and validated data, information, and knowledge for these models are among the challenges. Various solutions for these aspects have been investigated extensively for years, and many of them are application specific. Because this level of detail is beyond our scope, we will not elaborate the technical solutions in this book. However, note that the technologies both for AFs and SAFs are maturing and are a part of many simulation systems. More capable HW platforms that can support more complex solutions are offered in the market.

Finally, AFs can learn as they decide and act based on autonomous decisions. For example, let us assume that we initially set the observation angle of an AF as 90°. As this angle gets smaller, the target acquisition in the observed area gets quicker; however, the area not observed gets larger, which means a higher risk for receiving surprise attacks. This AF can readjust the angle of observation as it makes decisions and observes the results based on a given metric, such as the number of casualties. Of course, this also implies higher level of autonomy, which may not be desirable for some CAX planners.

9.5 CHALLENGES AND APPROACHES IN THE IMPLEMENTATION

9.5.1 Complex Systems and Fuzzy Trees

Complex systems and ill-defined information are one of the most influential factors in the decision-making process. When making a decision, one has to face the uncertainty of future events and the uncertainties that accompany transmission, transfer, and reception of information. Classic approaches to system modeling can hardly cope with complex systems and systems that involve uncertainty, for it is extremely difficult to find a global function or analytical structure for a nonlinear system. Fuzzy set theory[4] provides the mathematical frame of system modeling when knowledge and information about a system are incomplete and based on experience rather than systematic.

[4] The theory of fuzzy sets offers very good tools for presenting MCDM or multiobjective decision-making (MODM) problems where we deal with the fuzzy/inexact input data. To model these fuzzy numbers, membership functions or possibility clustering are used on case-by-case basis [LAI94].

Such problems are often encountered when applied to control, pattern recognition, or multicriteria decision making (MCDM) [SAV98].

Therefore, it is obvious that in real life many processes are closer to fuzzy processes than mathematically defined procedures. Many times, one gets the impression that in real life situations, problems are most efficiently solved by using the human factor, because computers have proved too complicated and impractical. As a consequence, there is an increasing need to improve computer capacities, so that we solve fuzzy decision-making problems. The latter can be defined as making decisions in a fuzzy environment.

A military organization is a typical hierarchical structure. It is relatively complex and involves many attributes. In addition, data on units are often incomplete, and force information is often either incorrect or even false. The following will have to be taken into consideration:

1. A military organization is a relatively complex and nonhomogeneous organization.
2. Many times, especially when a description of opposing forces is involved, a certain degree of uncertainty has to be accounted for.

In such cases, the description of a military organization takes the form of a fuzzy relational structure, that is, a fuzzy tree. Each node of the tree represents the status of a certain subunit. The status of a subunit can be described by using numerical values (men, weapons, ammunition, food, or water). In peacetime it is useful and necessary to use precise values. Such procedures are meaningless in actual war, given all the sources of uncertainty and decision-making complexity listed previously. Therefore, the status of momentarily usable and available systems can be characterized by the following linguistic units (used weapons, ammunition, men, and morale of the team):

- e (excellent); the number of active systems is about 100% ($B_e = 1.0$)
- g (good); the number of active systems is about 90% ($B_g = 0.9$)
- a (acceptable); the number of active systems is about 70% ($B_a = 0.7$)
- p (poor); the number of active systems is about 50% ($B_p = 0.5$)
- $ (missing); the number of active systems is about 0% ($B_\$ = 0.0$)

A fuzzy tree can be formed on this basis, defined over a fuzzy set X with a fuzzy labels nodes set L = {e, g, a, b, $} and characterized by a membership function $\mu_L{}^*(x, y)$. Sinusoid membership functions have proved the most suitable for this purpose [SAV98]:

$$\mu_L * (x_i, y_j) = \frac{1}{2} \cos \pi(x_i - y_j) + \frac{1}{2} \quad \text{for} \quad i, j = 1, \ldots\ldots 5$$

An important achievement in the development of the automation of command and control functions is the implementation of a system that collects and merges information from diverse sources into an appropriate representation of the tactical situation on the battlefield.

Such a system is often called a battle management system or combat decision aid system, which requires the following two functions [WAL87]:

- *Data fusion* collects information from a variety of sources to develop a possible overview of the military situation on the battle space
- *Decision support* carries out the creation of possible actions and a quantitative evaluation of alternatives on the basis of which a commander selects an appropriate alternative

One of the most important elements in the decision-making process is the estimation of the situation on the battle space.

Example 9.2

During combat simulation, we followed the number of destroyed systems and manning through the following phases of the combat: delay operations (d_1), battles on the firing line (d_2), defense operations (d_3), and counterattack (d_4). The above-mentioned phases have been compared with initial and reference situations. The initial situation shows the units before the beginning of the fight, whereas the reference situation represents units that were involved in the fight and are no longer capable of fighting. Each phases results in a different fuzzy tree. The distances between trees stand for changes in manning over time considering the initial situation. By monitoring these distances, a commander can easily follow the situation on the battlefield (Table 9.3).

The distances need to be normalized in compliance with a distance reference. The initial situation, which represents an excellently manned group, is ascribed a real value $B_e = 1.0$, whereas the reference situation, which represents badly manned group, is ascribed a real value $B_p = 0.5$. Normalized distances d^n are obtained by the following transformation:

$$d^n = B_e - (B_e - B_p/d_r)d, \quad 0 \leq d \leq d_r$$

where d_r is the distance between trees representing the initial and reference situations. Table 9.4 shows normalized distances between fuzzy trees.

Table 9.3: Tree-to tree distances.

Situation	Initial d_0	Phase 1 d_1	Phase 2 d_2	Phase 3 d_3	Phase 4 d_4	Reference d_r
Initial d_0	**0.00**	**0.63**	**3.64**	**8.40**	**11.85**	**22.55**

Table 9.4: Normalized distances between fuzzy trees.

Situation	Initial d_0	Phase 1 d_1	Phase 2 d_2	Phase 3 d_3	Phase 4 d_4	Reference d_r
Initial d_0	**1.00**	**0.98**	**0.92**	**0.81**	**0.74**	**0.50**

The results of the reasoning about the unit capacity at the certain situation can be represented by linguistic units (Table 9.5). With the increasing popularity, the use of hierarchical data structures and tree construction schemes, which is a special class of graphs, the problem of comparing two tree-like data structures arises. The distance between two fuzzy trees is calculated in terms of the minimum cost of deleting and inserting nodes and substituting node labels. The cost of those basic operations is calculated on the basis of fuzzy set theory. This makes it possible to handle not only problems in the deterministic environment but also problems that appear in the fuzzy environment.

There are two basic approaches to comparing tree structures:

1. The first approach is based on the use of the decomposition of complex structures into simple ones. In the tree structures, trees are divided into simpler subtrees applied to an adequate metrics [BOO73].
2. The second approach is based on the idea of so-called language transformations. A comparison of tree structures is based on the distance between the trees or an overall estimation of transformation of one tree into another [WAG74].

The second approach proved to be more appropriate in the areas of pattern recognition and support to decision making.

9.5.2 Minimalist Modeling Methodology

Because of already mentioned complexity of contemporary operations, the CAX planners have been coping with the complex modeling of the necessary systems of operation. Cobb [COB96] was one of the founders of the "minimalist" modeling methodology, which is a set of concepts that has evolved during

Table 9.5: Presentation of distances between fuzzy trees by linguistic units.

Dynamics of the combat	Initial t_0	Phase 1 t_1	Phase 2 t_2	Phase 3 t_3	Phase 4 t_4	Reference t_r
Situation of a unit	*e*	*E*	*g*	*G*	*A*	*p*
Comment	**excellent**	**excellent**	**good**	**Good**	**Acceptable**	**poor**

15 years of collaborative work with Dr. A. E. R. Woodcock on military, social, and ecological modeling, since 1980. This method consists of the following five fundamental ideas:

1. Structural simplicity
2. Behavioral simplicity
3. Mathematical elegance
4. Statistical analysis
5. Organic units

The purpose of minimalist modeling is to allow the analyst to focus attention and detail on one or two vital areas of interest, in which great modeling detail is used, while painting in all other areas of behavior and structure in broad strokes that are based on generic universal principles that apply to virtually all organizations, hierarchical or otherwise. The behavioral picture so created is then enhanced with statistical methods, so that in the end only the essential answer is given, and all other aspects are averaged out. In this way, minimalist modeling clears away much of the fog of detail and data that plague traditional combat modeling.

9.5.2.1 Structural Simplicity. In common with social and biological structures that exist throughout nature, military organizations and structures have a remarkable degree of physical and temporal coherence, and these structures seek to maintain that coherence even in the face of assault from predators. The models of social and military structures should reflect this dynamic as a fundamental self-organizing principle. When this is done properly, the shapes and structures of hierarchies emerge from the dynamic itself, just as the hexagonal shape of a beehive emerges from the efforts of bees to pack honey into a rigid structure made with walls of soft wax. The alternative, to prescribe each and every component of an organization, places an enormous burden on the modeler. To take a military example, massive databases must be created and maintained that specify the exact properties of every military unit, from fire team to army group, from artillery battery to naval squadron to recon team. The volume of data is huge; its reliability is questionable, and the construction effort is enormous. The minimalist method instead uses simple principles that apply universally. Continuing the military example, every military unit in the field, regardless of size or composition, deploys subunits to protect its flanks and rear areas. For another example, every modern organization attempts to limit the number of immediate subordinates that a supervisor has, so as to gain supervisory efficiency. When supervisory efficiency degrades because of accretion of subordinates, a new layer of command is inserted to restore the balance. This dynamic principle applies to every level of all hierarchical organizations, so they only need be encoded once for all levels.

9.5.2.2 Behavioral Simplicity. Like structural simplicity, behavioral simplicity is a general modeling principle. Rather than encode the decision-making behavior of each level of an organizational hierarchy as a separate algorithm, minimalist modeling attempts to use general principles of decision making that apply to all levels and sizes of organizations. Many business decisions, for example, scale to any size of corporation: merge, acquire, restructure, and reorganize. Similarly, military maneuvers also scale to any size of unit: frontal attack, flank attack, envelopment, ambush, and retreat.

9.5.2.3 Mathematical Elegance. At the core of all good modeling lies mathematical elegance. Whether the mathematics comes from deep but simple structural principles like symmetry and information, or from modern theories of the statistical behavior of nonlinear dynamical systems, the proper application of mathematics at critical points in a modeling exercise leads to stunning increments in efficiency and power. Wherever possible and appropriate, minimalist modeling attempts to use this natural weapon for slicing through apparent complexity and the chaos of details. As a simple but realistic example, when a component of a large war simulation involves several battles in which attrition, rather than maneuver, is the primary factor at work, then these battles can be reduced to an application of stochastic Lanchester equations without doing violence to the overall validity of the larger simulation. The gain in speed is huge; the loss in detail is trivial.

9.5.2.4 Statistical Analysis. Minimalist modeling takes advantage of a phenomenon of nature known as the Law of Large Numbers. Unlike deterministic modeling, stochastic modeling does not claim the ability to foretell the outcome of specific situations. Instead, repeated simulations are used along with well-known statistical methods to show trends in the outcome based on variable input data.

In deterministic modeling, a complete set of initial conditions is given, along with specific rules to define all actions to be taken when the model is run. Even though many good models of this variety are available, there are also many poor ones. In addition, a good deterministic model can be rendered useless by incomplete information regarding the initial conditions for a simulation, or by a lack of understanding as to how variables interact. Given good and complete data about the initial conditions, and a true understanding of how variables interact, these models may be able to predict the outcome of specific scenarios.

Minimalist modeling makes no such claims. Under no conditions, without regard to the validity of the input conditions or truth of the situational knowledge provided by the user, can a minimalist model predict the specific outcome of an individual situation. What it can do is provide the user with insights as to how varying the conditions will affect the outcome of battles. These effects remain unknown until the output of repeated simulations is massaged by statistical software.

The validity of the minutiae in simulation scenarios is of secondary importance. Over a large number of runs, while experimentally varying critical parameters, statistical truths will make themselves known. So, in preparing a simulation, the analyst should portray the initial conditions as accurately as possible, but not become overly concerned if some conditions are not known.

It must be said that the ability of the analyst to glean useful information from these simulations is in direct proportion to the analyst's understanding of the statistical analysis. Graphic presentations made by a simulation should be used as a tool for testing the face validity of a simulation scenario. The output from the simulation should be treated as statistical data and reported as such. Whereas a determinist might say, "If you build your embassy there, a car-bomb will destroy it," minimalists will only say, "The survival probability of embassy personnel improves in proportion to the square of the distance of the embassy to the street."

9.5.2.5 ***Organic Units.*** Organizations are composed of myriad parts. Minimalist modeling does not attempt to describe an organization in terms of these parts, because to do so leads directly into data base madness. Instead, the minimalist description of an organization is functional. Here are two examples of how this is done. The first example is governmental, and the second is military.

In the governmental example, instead of listing the individual agencies that exist at each level of government, with their individual organization charts, budgets, and legal responsibilities, we rely on a graph of each government's ability to regulate the transactions of all people and institutions within its boundaries as a function of certain key variables of interest (e.g., land area, size population, size of economy, travel and communication speed, and technological development). Each level of government is treated as an organic entity, without regard for the specific details of methods and means by which its goals are attained. Because these functions have very few parameters, the entire description compresses down into an efficient format. The evolution and pathologies of government can be modeled and studied in this minimalist way without exhaustive descriptions of myriad individual agencies.

In the military example, instead of listing the range, accuracy, and number of each type of weapon that a unit has, we rely on a single graph of the unit's firepower as a function of range. Instead of listing the range, accuracy, and number of each type of sensor that a unit has, we rely on a graph of detection probability as a function of distance. Because each function has only one or two parameters, the entire description compresses down into an extremely efficient format. Compared with standard military unit descriptions, the compression factor is on the order of two, three, or even four orders of magnitude.

This functional approach works well because real military units are constructed as organic, functional entities. Insofar as humanly possible, weapons and sensors are selected to cover a range of capabilities, so that the unit is never left defenseless or blind any specific range. The specific weapon or sensor is not important, but the coverage of all weapons and sensors is. Minimalist modeling describes the coverage, not its detailed implementation with individual weapons and sensors.

Exactly that characteristic of minimalistic modeling has changed the presentation of CAX analysis data. All data have been explained as a quantitative state of consumed material and losses of unit, and not qualitative ratios among data. With the help of multivariate analysis, it is possible to define variables that have the biggest impact on execution of CRO, and at the same time to improve quality of the research processes [MAR05].

9.6 COMBAT MODEL DATA

A data base is organized form of data management, which allows quick access to information. For war gaming, it is necessary to prepare some different data bases that will be described.

9.6.1 Organizational Data

Organizational data consists of a hierarchy of units and all details about their organization, for example, the number of battalions in a certain brigade or the number of squads in particular platoons.

9.6.2 Equipment, Weapons, and Ammunition Data

Most data that are needed for war gaming could be classified in data about equipment, weapons, and ammunition. Vehicles, sensors, and launchers are part of equipment, but cannons, mortars, and infantry weapons, for example, are an integral part of weapons. Data bases are interrelated because singular weaponry systems use ammunition at the same time they are part of equipment. Thus, data structures are complex, which leads to the need for an appropriate system for data management. Obviously, there are interrelations among data about equipment, weapons, and ammunition, and because of that there are *relations data bases*.

Equipment data depends on the type of equipment. For example, vehicle equipment consists of data about maximum speed of vehicle, consumption of fuel, type of weapons on the vehicle, action range, and so on. Weapons data consists of data about minimum and maximum range of fire, cadence, and accuracy of fire. Ammunition data consists of data about lethality range, weight, and penetration. Data depend on the model that has been used in the war gaming.

9.6.3 Terrain Data

A map of the terrain could be digitized or prepared in another way. Digitalization of a map has been executed in different layers, with squares or hexagons. Each square or hexagon has attributes, for example, height above the sea and vegetation. A particular layer is dedicated to the waters (rivers, channels, lakes, coast). Rivers have an attributes such as depth and width, inclination of embankment, and possibility to lay down pontoon. There is a layer with the land communications (roads and railway lines), which have an attributes about road length, width, and class of the road. With all above-mentioned data, the user creates an *intelligent map* [SAV00]. The terrain data have been used in war gaming for definition of vehicle speed limits, visibility, accessibility, objects, obstacles, and so on.

9.6.4 Environmental Data

Environmental data consist of data about seasons (summer, winter, and rainy), weather (temperature, rain, snow, and ice), daytime (day or night) and other data that have an impact from environment on the game.

Example 9.3

Combat simulation model HORUS has a number of data (Figure 9.10), which could be classified in the following data groups:

- *Configuration*: data about computers set up and dynamic simulation parameters
- *Terrain*: static terrain data
- *Situation*: dynamic situation on the battle space
- *Parameters*: static parameters in simulation (ammunition, vehicles, and weapons.)
- *Recording*: changes of the data group situation, during simulation. It has been used for the presentation of the simulation, for time jumps, and analysis.

9.7 VERIFICATION AND VALIDATION OF COMBAT MODELS

The model is a projection of the real-world situation, where the user expects to have a forecast for the real-world system behavior. With the available data, it is possible to compare statistically the behavior of the model and real-world system. Statistical analyses have an important role in validation of models. These analyses consist of analysis of deviation, verification of a cluster analysis, regression, and correlation analysis. The research of the military systems is suffering because of lack of historical data. That fact has impact on validation of combat models, where the main roles have military experts with their perception and understanding.

Verification of the combat model validity has the following basic procedures:

1. *General validity*: experts can give their opinion about certain model, particularly about their satisfaction and model behavior.
2. *Sensitive validity*: validation of variables (sample) and parameters (entire population) in comparison with changes of the real-world system.
3. *Validity of hypothesis*: study of relations in the model need to meet similar relations in the real-world system [SAV00].

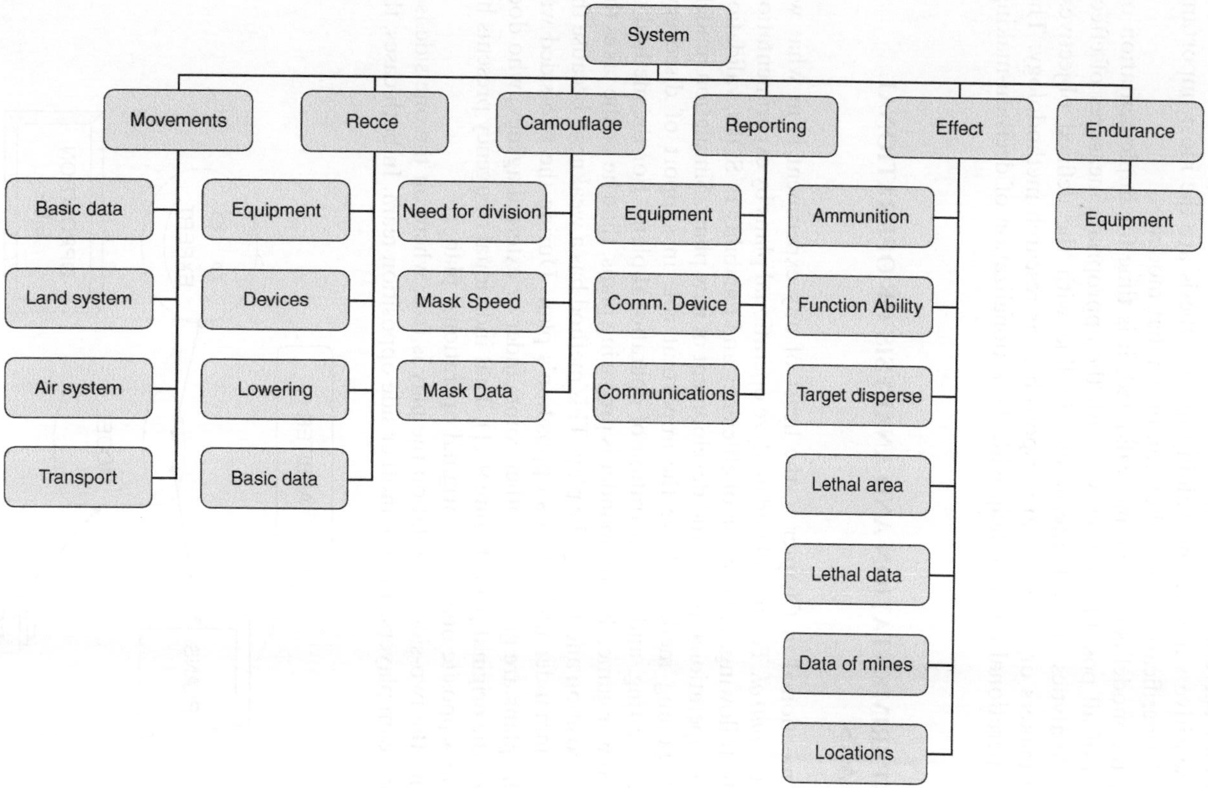

Figure 9.10: Data structure in system HORUS.

Military systems cause casualties and damage, because of that it is difficult to have many reiterations or reproductions. At best, it is possible to use data from the tests and exercises, where users are using artificial targets. The general validity, sensitive validity, and validity of hypothesis are the most important methods for verification and validation of combat models.

When the model is created and validated, it is time for implementation or validation of all possible alternatives with the appropriate measures of effectiveness. Analytics suggest alternatives in line with the defined objectives. Decision makers or executors are responsible for research methodology. The military operational research help with the rationalization of decision-making process.

9.8 EXPERIMENTATION AND ANALYSIS OF OPERATIONAL PLANS

As was mentioned in Chapter 6, the heart of any experiment is in what we attempt to *control*. To control and analyze operational plans by experimentation means the following: verification of efficiency and reliability of SOPs, validation of existing operational plans, and development of new plans. Simulation models and war gaming methods have the important role in support of discovery, hypothesis testing, and demonstration of operational plans. For the method of one- sided war game, the commander is preparing plans, than he or she takes the role of opposition and testing the plan. This method has a weakness, because the playing commander understands opposition's plan. During the two-sided war game, the plans are given to another commander or external agency, who does not know the original plan (Figure 9.11). That individual or agency presents his plan as an opposite one to the original operational plan.

During the two-sided simulation the play is free, where on the one side is a customer team players, and on another side opposition team. In both cases, the

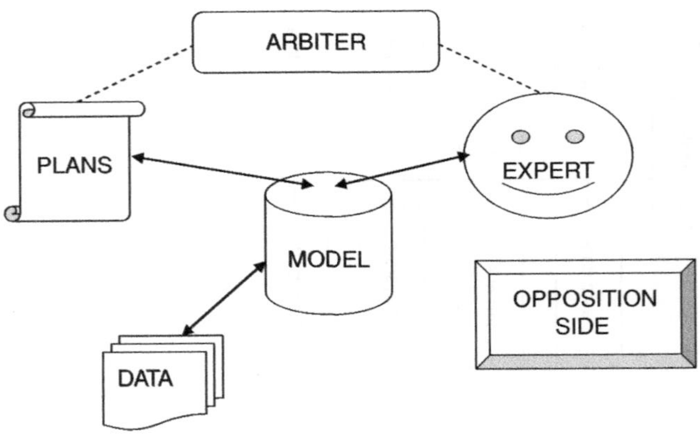

Figure 9.11: Set up for the analysis of operational plan.

operational plans a refresher to the noninvolved players, which were not involved in the development of plans. Many gaps could be found during the experiment, which were not recognized by the planners. The opposition team can play certain alternatives, which nobody predicted before. With the repetition of the games, the operational plans can properly validate, update, develop, and improve up to the highest desirable level.

9.9 REVIEW QUESTIONS

9.1 What does combat modeling allow?

9.2 Define synthetic environment.

9.3 How are the key data sources of environment defined?

9.4 What is the role of SEDRIS data coding standard?

9.5 What are the main differences between deterministic and stochastic models?

9.6 Why are Lanchester's equations important for simulation of real combat?

9.7 What is significant about stochastic simulations?

9.8 Explain the method for quantification of nonkinetic activities.

9.9 Explain the groups of automatic forces objects.

9.10 What is a contribution of the minimalist modeling methodology to the modeling?

9.11 Which basic procedures have been used for verification of the combat model validity?

10

COMPUTER-ASSISTED EXERCISE SUPPORT TOOLS

Let us remember our definition of (CAX). A CAX is a computer post exercise (CPX) where electronic means are used for the following:

- To immerse the training audience (TA) in an environment as realistic as possible
- To help the exercise planning group (EPG) and the exercise control (EXCON) staff to control the exercise process (EP) so that it achieves the objectives as effectively as possible

Therefore, CAX support is more than setting and running a military simulation system. CAX tools should be involved in all stages of an EP explained in Chapter 8 to automate the processes, to reduce the duplication of work, to enhance the exercise environment, and to ensure that the EP flows towards the objectives. CAX tools can be categorized into four classes:

- ***Simulation systems and ancillary tools***: These tools are simulation systems and the software needed to prepare and to run simulation systems (e.g., database preparation tools, user interfaces, etc.).
- ***Exercise planning and management tools***: These tools can be used for the automation of processes, information management, and information exchange for the preparation of the exercise specification (EXSPEC) and exercise plan (EXPLAN) documents and the products related to these

documents. They can help the preparation of scenario as well as the main event lists (MELs) and main incidents lists (MILs). They can also have interfaces for the CAX tools that fall in the other categories. Through these interfaces, the data collected during the specification and planning stages can be directly fed into simulations.

- ***Mediation ware between simulation and command and control (C2) operational planning (OP) tools***: Simulation systems should be transparent to TA. Especially primary TA (PTA) should only use C2 systems that can be available during an operation. Therefore, interfaces between the simulation software and C2 systems are needed. Similar interfaces are also required for the OP tools because they also need data related to the exercise scenario and the current situation.
- ***Experimentation and analysis tools***: These tools are the programs for designing and managing experiments by using CAX data and for compiling and presenting the data collected by the simulation system as well as deriving information from these data.

10.1 MILITARY CONSTRUCTIVE SIMULATIONS AND ANCILLARY TOOLS

Military simulations are categorized as live, virtual, and constructive. See Chapter 1 for the details. Among these classes, military constructive simulations are often the only class of military simulation systems used in a computer-assisted exercise. As the technology matures, live and virtual simulations federated with constructive simulations are more extensively used in CAXs. Before elaborating the topic, we first would like to clarify several issues related to this classification of live, virtual, and constructive.

First, there has been a debate about the scope of live simulations. In Chapter 1, a live simulation is defined as a simulation that involves real people who operate real systems. In live simulations, the interactions between the real systems and people are simulated. A good example for this is that two real aircrafts are dog fighting, and a simulation system decides whether the engagements are successful. Similarly, ground units use laser emitters and sensors to decide the results of engagements. Many people believe that when a command and control center joins an exercise by using the real C2 devices connected to a synthetic environment, it is also called a live simulation because C2 devices and the people using them are real. Although some experts do not agree with this, we believe that this is correct interpretation. Therefore, we can claim that the CAX community has been using live simulations for a long time because many state-of-the-art centers can feed the C2 devices based on the results from the simulation systems.

Live simulations are also often mixed with live exercises. In essence, a live exercise is a live simulation, although it is not a computer simulation. However, the uses of live simulations are not limited to live exercises. In reality, a live exercise can be supported also by a constructive simulation system, and a command post exercise can be supported by a live simulation. For example, while one platoons is in the field running a live exercise, the other platoons of the company can be simulated in a constructive simulation system. Similarly, live simulation systems like real command centers can be used in a CAX.

Virtual simulations are often perceived limited to things like aircraft, ship, and tank simulators. Virtual simulations can also be used to create realistic environments for command posts. Computer graphics, motion, fog, and light generators can be used to create an environment that looks like a battlefield, and leaders can be asked to make their decisions in such an environment. Moreover, virtual simulations can be used to create products like video streams from unmanned aerial vehicles or two-dimensional (2-D) three-dimensional (3-D) imagery from satellites and/or aircrafts. Especially when virtual simulations can interact with constructive simulations by using distributed simulation technologies, they become extremely useful for a CAX at any level.

We focus on constructive simulation systems in this chapter. Some of the high-resolution constructive simulation systems also have embedded virtual reality tools, generally in the form of very realistic imagery and video. We also explain live virtual constructive (LVC) federations. We believe that LVC federations will be more extensively used in CAXs.

10.1.1 High-Resolution Constructive Simulations

High-resolution constructive simulations are typically for tactical levels starting from a single troop up to several brigades. The terrain, weather, and entities are simulated detailed in these models. Each weapon, individual soldier, and combat system can be a simulated entity. Terrain modeling can be as detailed as centimeters, leaves of trees, and furniture in a room. Engagements are modeled typically between entities. Computations can be done for each single bullet shot by a troop.

As the level of detail increases, the more detailed data and the higher hardware capacities (i.e., memory and computational power) are required. Hardware capacities introduce limits on the size of simulation (i.e., the number of entities and the size of the simulation area). Therefore, there is a tradeoff between the level of detail and the size of a simulation.

As the hardware capacities increase, the limitations on the size of simulation disappears. For example, a typical play box for a high resolution constructive simulation system used to be 200 kilometers × 200 kilometers a decade ago. Nowadays, there are high resolution constructive simulation systems that can simulate as many as 50,000 entities in an area as large as 2000 kilometers × 2000 kilometers.

Apart from the hardware constraint, the other factors like the level of planning and the number of operators also affect the selection between a high-resolution or highly aggregated simulations. The higher the level of detail a model has, the more manpower is required to run the model because more details are needed in the commands. Aggregation and deaggregation capabilities have been introduced to tackle this challenge. In many high-resolution simulation, it is possible to aggregate entities into units and to command units instead of each individual entity. However, still much more detail is required for the commands given to the aggregated units in high-resolution simulations compared with the units in highly aggregated simulation systems. Therefore, although high-resolution constructive simulation systems can provide higher fidelity, they may still be not appropriate to run exercises in echelons higher than corps level.

In Table 10.1, some legacy high-resolution constructive simulations and some state-of-the-art systems are listed. The list is far from being exhaustive. Our aim is to provide a set of examples for the rest of this section.

Let us start with terrain representation in high-resolution models. Note that terrain modeling is explained in detail in Chapter 9. In many high-resolution constructive simulation systems, geographic data are generated from DTED elevation data. DTED-1, which provides elevation data in every 10 meters, is very common for this category. Similar to DTED, VMAP is commonly used for vector data. Although many simulation systems use DTED and VMAP to generate the environment data, still almost every constructive simulation system uses some proprietary formats for the terrain and environment data. For example, brigade battalion simulation (BBS), Janus, JCATS, and OneSAF are all U.S. high-resolution constructive simulation systems. They use some proprietary formats and represent the terrain data base different from each other's.

Play box limitations used to be more stringent for high-resolution simulations. For example, JCATS play box used to be limited to 200 kilometers × 200 kilometers. Now it is 2000 kilometers × 2000 kilometers which is much larger than an area needed for an exercise in tactical level. Therefore, when the model is used in the proper level, for example not to simulate intercontinental movement plans, almost none of the state-of-the-art high-resolution systems introduce a play box size limitation. Similar to play box size, the numbers of entities that can be simulated effectively in a state-of-the-art high-resolution system do not introduce stringent constraints when it is used in the proper level.

Virtual simulation capabilities, especially graphics, embedded into high-resolution simulation systems are common with the state-of-the-art systems. Here, we can classify high-resolution simulations into two more categories. In one class, we have simulation systems that focus on unit (platoon, company, battalion, brigade, etc.) tactics. The other class focuses on individual (single troop or squad leader) tactics. The former class has only 3-D viewers, which do not provide much fidelity. The later class has more realistic and enhanced virtual reality capabilities. VBS2 falls in this category. Other serious commercial games can be used with the same purpose. The graphics generated by the tools like VBS2 is so realistic that they can be used as a video stream coming from an unmanned aerial vehicle or a satellite or aerial imagery. Therefore, they are useful not only for small unit or individual training but also for higher level tactical and operational exercises.

Table 10.1: High-resolution constructive simulation systems.

Name/Service/Source Nation	Terrain	Play Box in Kilometers	Entities	Virtual	Automated
Virtual Battle Space 2 (VBS2)/Joint/ Australia	Rapid terrain generation from DTEP, shape and imagery files. It can import three-dimensional (3-D) models (buildings, vegetation, etc) from 3DS or OpenFlight.	Up to 350 × 350 for DTED-1. Up to 120 × 120 for DTED-2. Up to 40 × 40 for DTED-3.	More than 1000 artificial intelligence (AI) entities. 120-200 human players.	Real-time rendering and highly accurate 3-D representations of objects, forces, and terrain.	Scripted semiautomated behavior.
Gefects Simulation (GESI) or Simulation four Rahmenubungen (SIRA)/Army with air and maritime entities/Germany	Uses TerraVista to read many data formats, such as digital terrain elevation data (DTED), digital height model (DHM), digital feature analysis data (DFAD), authoritative topographic cartographic information system (ATKIS), and geographic tagged image file format (GEOTIFF).	Up to 2000 × 2000.	Up to 32,000.	3-D view of terrain and entities.	Artificial intelligence to create autonomous forces from selected entities.
Korp-Rahmenmodell (KORA)/Army/ Germany	Interface formats are DTEF, DFAD, and GEOTIFF.	Up to 1000 × 1000.	Not limited.	No.	Behavior agents.
SCIPIO/Army/France	DTED-1, Vector map 0 (VMAP0)/VMAP1, Georeferenced maps or photos.	No limitations. Currently used in exercises with a play box of 2000 × 2000.	No hard limitations. Several thousand entities or units.	No.	Semiautomated forces.

(*Continued*)

Table 10.1. Continued

Name/Service/Source Nation	Terrain	Play Box in Kilometers	Entities	Virtual	Automated
SCEPTRE/Airforce/France	DTED-1, VMAP0/VMAP1, Georeferenced maps, or photos.	No limitations Currently used in exercises with a play box of 2000 × 2000.	No hard limitations. More than 1000 air mission a day.	No.	Semiautomated forces.
CATS-TCT/Army/Sweden	DTED, VMAP, georeferenced maps, or photos.	No limitations. Play box enough to cover a brigade level exercise.	No hard limitations. As many entities as can be in a brigade.	3-D viewer.	No.
DEHOS/Navy/Turkey	High resolution from VMAP and shape data.	5 million square nautical miles.	No hard limitations. More than 1000 naval entities.	No.	No.
Joint Conflict and Tactical Simulation (JCATS)/Joint/US	Terrain generation from DTED, shape and imagery files.	4000 × 4000.	No hard limitations, about 100,000 entities. Up to 10 sides.	No	No.
One Semiautomated Forces (OneSAF)/Joint/US	Very high resolution (1/12,500). High-resolution buildings (elevator shaft, balcony, stair, etc.).	500 × 500.	Up to 25 sides. Entities up to brigade level. In high resolution upto 500 entities. In low resolution up to 5000 entities.	3-D viewer.	Enhanced semiautomated force behaviors.

State-of-the-art high-resolution constructive simulations typically have also automated forces capability, generally in the semiautomated force notion. We believe that automated forces will be used more commonly in constructive simulations.

Finally, many high-resolution simulations are developed with one service in focus. Although many are claimed as joint, only few high-resolution simulations are truly joint (i.e., in equal distance to any service).

10.1.2 Highly Aggregated Constructive Simulations

Examples for the highly aggregated constructive simulation systems are listed in Table 10.2, which is again far from being exhaustive.

The first major difference visible to users between high-resolution and highly aggregated simulation systems is the representation of the terrain and environment. In highly aggregated systems, the play box is tessellated with either hexagons or squares, and each of these hexagons or squares represents the following:

- Terrain characteristics (i.e., forest, ocean, desert, etc.)
- Mobility characteristics (i.e., good, bad, no mobility, etc.)
- Altitude or depth

Moreover, the sides of these hexagons or squares are used to introduce obstacles like rivers, tank ditches, shores, minefields, and so on. For example, a river that can be an obstacle for the unit mobility must follow the edges of these geometric shapes. This approach may not look very realistic, and sometimes the results from simulation do not match with the maps and the data in C2 systems. For example, the real location of the river may be several kilometers different from a hexagon edge. Because the model uses the hexagon edge as an obstacle, a unit may stuck somewhere that does not look realistic.

The size of hexagons/squares can change from one model to another or even from one execution to the other of the same model. An approximately 3-kilometer radius is typical for the hexagons as shown in Figure 10.1. This generally does not provide the required level of fidelity. Many highly aggregated simulation systems are legacy systems with millions of lines of code, and many important algorithms like detection, engagement, movement, and attrition are based on the terrain representation. Therefore, things like reducing hexagon size considerably (e.g., 1-kilometer radius), are not quick fixes for them, although it looks like at the first glance.

Hexagons have some advantages over squares. If all the hexagons are the same in size, then the distance from a hexagon to all its neighbors is the same. This layout is different from squares, where the distance to the diagonal neighbors are longer than the distance to the vertical or horizontal neighbors. This introduces ease in computations when hexagons are used, and therefore many highly aggregated simulation systems model terrain by using hexagons.

Table 10.2: Highly aggregated constructive simulation systems.

Name/Service/Sorce Nation	Terrain	Play Box in Kilometers	Sides/Units	Automated
WAGRAM/Army/France	DTED-1, VMAP0/VMAP1, georeferenced maps, or photos.	No limitations Currently used in exercises with a play box of 2000 × 2000.	No hard limitations. Several thousand entities or units.	Semiautomated forces.
ORQUE/Navy/France	DTED-1, VMAP0/VMAP1, Georeferenced maps, or photos.	No limitations Currently used in exercises with a play box of 2000 × 2000.	No hard limitations. Several thousand entities or units.	Semiautomated forces.
Simulations Modell fur ubungen Operativer Fuhrung (SIMOF)/Army-Air/Germany	Interface formats are DFAD and GEOTIFF.	Up to 2500 × 2000.	Not limited.	No.
Air Land Interactive Conflict Evaluation (ALICE)/Airforce/Germany	DTED, DFAD, vector, and scanned maps.	4000 × 4000.	Not limited.	No.
CATS-TYR/Joint/Sweden	DTED, VMAP, georeferenced maps, or photos.	No limitations. Play box enough to cover a corps level exercise.	No hard limitations. As many entities as can be in a brigade.	No.
Joint Operational Command and Staff Training System (JOCASTS)/Joint/UK	Hexagons. The size of hexagons can be changed.	Corps size exercises.	No hard limitations.	No.
Joint Theater Level Simulation (JTLS)/Joint/US	Hexagons. The size of hexagons can be changed. The side length can be as short as 1 kilometer. However, when the side length is less than 3 kilometers, the performance of model depends on the scenario and number of units in the scenario.	4000 × 4000.	Up to 10 sides. As many as 10,000 units (no hard limit).	No.

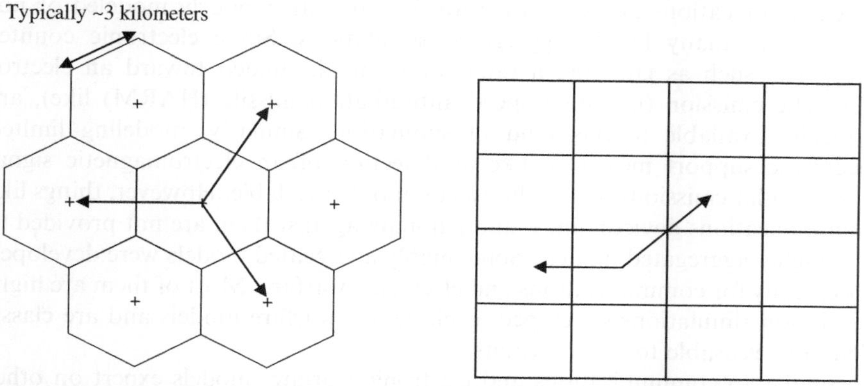

Figure 10.1: Terrain representations in highly aggregated constructive simulations.

The major advantage gained in the expense of lower fidelity is that highly aggregated simulations can fit larger simulation size (i.e., larger play box and larger number of units). They are typically for theater- or intertheater-level simulations. The play box can be larger than 4000 kilometers × 4000 kilometers. Thousands of units made up of hundreds of thousands entities can be simulated with high simulation speeds like 100:1 (i.e., each real second is equal to 100 simulation seconds).

Of course, this speed is gained not only in the expense of lower fidelity in terrain modeling but also in the expense of attrition and movement modeling. Highly aggregated simulations use the Lanchester equations for computing the attrition especially for the land battles between units. They may have stochastic processes to compute the results of things like indirect fire, air-to-air, and surface-to-air engagements. Still, these processes are lower in fidelity compared with the high-resolution systems. However, this disadvantage comes with an advantage: The operators need to enter less detail when they pass their commands to the model. Also, they fit better to the level of abstraction and procedures of operational and higher level headquarters. Therefore, they are preferred in operational or higher level exercises.

However, the new generation warfare changes the requirements. The term "strategic corporal" is heard many times in recent operations. Urban warfare, peacekeeping, and peace support operations require tactical details even at the operational level. Also, some types of operations may require tactical considerations during the planning in operational level. River crossing, amphibious operations, long distance recce, and ambush are some examples for this kind of operations. Therefore, many tactical-level details and modeling have been added into many highly aggregated simulations. The bottom line is that high-resolution systems are becoming more capable in simulating larger scenarios, and highly aggregated simulations are gaining more fidelity; therefore, we can say that the gap between these two categories has been closing. However, there is still a distinction between them, and we believe that it will stay still some time.

Communications and electronic warfare are either poorly modeled or not modeled in many highly aggregated simulations. Some electronic counter measures, such as jamming a radar or a missile guided toward an electromagnetic emission (i.e., high speed antiradiation missile (HARM) like), are typically available in this kind of simulation. Similarly, modeling limited electronic support measures, like as detection of an electromagnetic signal (e.g., a radar emission) is also almost commonly available. However, things like communications devices, links, and jamming against them are not provided in any highly aggregated models. Some highly aggregated models were developed specifically for communications and electronic warfare. Most of them are high-resolution simulations developed as electronic warfare models and are classified/not releasable to other nations.

Similar to communications and electronic warfare, models expert on other specific topics can be found. Both highly aggregated and high-resolution logistics, space, air defense, theater level missile defense, and chemical, biological, radiological, and nuclear (CBRN) simulation systems are available. Please also note that not all of the generic models are joint. There are a few joint models and many service models.

10.1.3 Constructive Simulations for Nonkinetic Warfare

Traditional military simulation technologies focus on kinetic warfare, where forces sustain, move, and engage each other in a theater. In essence, these systems are designed to compute attrition and movement of the units or entities in a battlefield. That has been an intensive effort to model realistically these kinetic aspects of the earlier generations of warfare for decades, and we can claim that the available technology is pretty mature in this respect. New generation warfare requires us leaving our comfort zone into a less known and more complex domain. The commander's plan is based not only on military tactics but also on political, economic, and social circumstances. Military leaders deal with local populations and must consider the issues like their security, welfare, and cultures. These issues are not within the scope of the known live, virtual, and constructive simulations, or at least were not initially. We need new types of simulations that can model human behavior, especially the behavior of social entities (i.e., societies, leaders, and organizations), as well as the interactions and interdependencies of these entities. The models have to predict the results from the interactions among different cultures in harsh conditions. They have to reflect accurately the characteristics of ethnic and religious groups, which change temporarily and geographically. These differences are nuances and from one nation to the other. Even a small nation of several million may have a huge diversity in the number of cultures and their differences. Therefore, the challenge is not only building realistic models for non kinetic warfare but also maintaining them and their data to ensure up-to-date information is used.

One approach to model nonkinetic aspects is to represent a society as a network of nodes. In this network, each node is a social, economic, or political entity, such as a leader, institution, organization, corporation, facility, or procedure. The network models the interdependencies of these nodes. When something affects one node, this effect is not only given to the initial node but also propagates through the network and affects the other nodes based on the characteristics of the links among the nodes. At the end, social, economical, and political impacts of the initial affect are observed in the overall network. This model fits very well to the effects-based approach to operations. To simulate nonkinetic warfare also helps to support operations and exercises where headquarters are organized according to the comprehensive approach.

One good example that follows this network approach is synthetic environment for analysis and simulations (SEAS) technique developed by Purdue University. It was designed initially to simulate strategic plans for commercial companies then was adapted for modeling nonkinetic aspects during a war. SEAS already has data bases about many nations. Iraq and Afghanistan are the most highly developed and complex models. Each has about 5 million individual nodes at the time we write this book.

The joint nonkinetic effects model (***JNEM***) is another example from the United States for the simulation of nonkinetic warfare. It has modules to interact with the other constructive simulations to receive inputs about the military activities and to pass back the results of those activities in the civilian population. Military operations, diplomacy, humanitarian efforts, civil construction work, and the other actions that a commander can take affect the public view, and JNEM simulates that.

Operations other than war have different requirements from the conventional war. Crowd control, refugee camps, disaster relief operations, crises and consequence management, border control, defense against terrorism, patrolling and check points during a peace support operation, embargo, and boarding a suspected ship are some examples that a typical combat model could not simulate less than two decades ago. Nowadays, many military constructive simulations including legacy simulations can model some or all of these kinds of operations because it is relatively easier to adapt the combat models for these purposes compared with the other nonkinetic war aspects. Simulation systems are developed from scratch specifically for these purposes.

10.1.4 Federations

As distributed simulation technologies mature, federations are used more frequently in support of CAXs. The earliest endeavor in this direction is joint training confederation (JTC) based on aggregate-level simulation protocol (ALSP). The JTC is a U.S. Defense Advanced Research Projects Agency (DARPA); this project started with two models in 1992. In 1997 there were 12 models in JTC, which include the following:

- Corps Battle Simulation (CBS) for Army
- Air Warfare Simulation (AWSIM) for Air Force
- Research, Evaluation, and System Analysis (RESA) for Navy

- Marine air ground task force tactical warfare simulation (MTWS) for Marines
- Combat Service Support Training Simulation System (CSSTSS) for logistics
- Joint Electronic Combat Electronic Warfare Simulation (JECEWSI) for electronic warfare
- Tactical simulation (TACSIM) for intelligence
- Portable space model (PSM) for space

JTCs were used in several exercises. It was a complex system that crashed more often than a typical constructive simulation. It was also more expensive to set up and run. ALSP and distributed interactive simulation (DIS) inspired high-level architecture (HLA) as we already explained in Chapter 5. With HLA, JTC has also evolved to a new federation called joint land component constructive training capability (JLCTC). Some models retire from JTC when it becomes JLCTC. Some other models, such as CBS, are also retiring from JLCTC when we are writing this book. However, new models, such as WARSIM, OneSAF, logistic federate (LOGFED) and JNEM, are either planned to be included or already being integrated.

In JTC, the aim was to federate service models, such as CBS, AWSIM, and RESA, as well as models designed for specific purposes, like CSSTSS, JECEWSI, and TACSIM. All these models are constructive and designed for the same level of training. A JTC-like approach serves mainly for interoperability and reusability purposes.

HLA technology can also be used for building multiresolution federations made up of both highly aggregated and high-resolution constructive simulations. One good example for that is joint multiresolution model (JMRM), which is also the basis for North Atlantic Treaty Organization (NATO) Training Federation (NTF). At this time, two main models in NTF are JTLS and JCATS. Therefore, NTF can simulate both at tactical and operational levels.

Federations of live virtual constructive (LVC) simulations are also available. Joint live virtual constructive (JLVC) is an example for that. JLVC and JMRM have common constructive simulation systems like JCATS. Still, they have different federation object models (FOMs). It is also possible to design a reference FOM and a reference federation with this FOM, which can be both LVC and multiresolution. For a specific exercise, a selected part of this federation can be used. HLA 1516-2008 (i.e., HLA Evolved), provides new concepts and architectures, such as modular FOMs that support this approach better. NTF design studies pursue this approach. Another example reference federation that falls in this category is Persistent Partnership for Peace Simulation Network (P2SN) Federation, which has many federates including, tactical command trainer (CATS-TCT), wargaming system for joint operations and operations other than war (CATS-TYR), JCATS, VBS2, GESI, OneSAF, and advanced simulation combat operations trainer (ASCOT).

Distributed simulations introduce many advantages already explained in Chapter 5. As the technology matures, federations become more feasible. Therefore, we believe that they will be used more often in computer-assisted exercises.

10.1.5 Ancillary Tools

Ancillary systems reduce the overall cost of using constructive simulation systems during CAXs, which mainly include the tools that support the data base preparation and review processes for the simulation and increase the efficiency of the procedures for the interaction between the EXCON and the simulation. Please note that the interaction between the TA and the simulation is different from the interaction between the EXCON and the simulation. That will be elaborated in the section about mediation ware.

Data base building is perhaps the most critical and time-consuming activity in a CAX preparation, which is the number one prerequisite for the success and realism of the simulation. The data required for CAXs are explained in Chapter 9.6. Various parties are involved in the preparation of those data, which include data providers, data base manager, and the data base management team coordinator. Data providers include the training audience who provide the data about own forces. A good set of tools is needed to support the process, to prevent duplication of work, to ensure the quality of data, and to avoid unauthorized exposition of the collected data.

Almost every simulation system comes with a data base editor. It is also very common to develop additional data base editors that fit to the data base building process. For example, JTLS has data base development software called data base development system (DDS). That suffices to build a JTLS data base from scratch. However, NATO developed also an ancillary data base preparation system called ORBAT Editor, which is used for collecting the data from data providers. ORBAT Editor provides a set of features that support the data base preparation and review process for JTLS. These features include tools to review and update as follows:

- The unit data such as the name of the unit, the combat systems owned by the unit, associated targets, the other units that have the same characteristics
- Command structure
- Support structure

ORBAT Editor allows multiple data providers to collect data for their forces without exposing unauthorized data and interacting with the other data providers. The data collected by ORBAT Editor are merged into a single data base and are edited and managed by using DDS by CAX data base managers. ORBAT Editor is used not only by NATO but also by some NATO nations, and there are many other similar tools with similar purposes for almost every constructive simulation system.

An important part of a CAX data base is terrain data, which also requires tools to manage. As it is depicted in Table 10.1, most simulations have proprietary terrain data structures, but they also have tools to generate them from standard data formats such as DTED and VMAP. Similar tools are also available for highly aggregated simulation systems where terrain is represented by hexagons or large squares. Terrain editor (TEREDIT) for JCATS, terrain modification unit (TMU) for JTLS, and GlobalMapper for VBS2 are examples for this kind of ancillary software.

Apart from ancillary tools for data base preparation, there are tools for supporting the directing staff and CAX operators to interact with a simulation system easily. They can be used to fetch data from simulation, merge it with other data, and present to the directing staff in a more operational format. They can also be used to convert the operational plans to the simulation commands.

10.2 PLANNING AND MANAGEMENT TOOLS

CAX support tools may support an EP starting at the EXSPEC stage. A software tool can provide a structured way to derive the training objectives, missions, and operational tasks based on current documents such as training/exercise programs, training/exercise directives and guides, lessons learned databases, the statements by the officer specifying the exercise (OSE), the officer conducting the exercise (OCE), and the PTA Commander. Then this tool may compare the requirements with the capabilities of the available simulation systems, and it can even figure out the required resources, (e.g., man power, CIS infrastructure, the cost of simulation, etc.) according to some additional parameters, such as the level of exercise and the size and the number of component commands. Because the number of available simulation systems is often small and it is easy to decide on the appropriate system, this tool can be more useful in determining weaknesses of a simulation system and approximate cost of using them for a certain exercise.

After the EXSPEC, OCE issues the exercise planning guidance. The next key milestone in EP is the promulgation of the EXPLAN. In the preparation of EXPLAN, many exercise support products including the following are prepared:

- Road to crises
- Master events list (MEL)/master incidents list (MIL)
- Operational planning process documents
- Political resolutions
- Force activation/transfer of authority messages
- Intelligence products
- Friendly order of battle (FOB) Enemy order of battle (EOB) databases
- Geographical information system databases
- Operational plans

Many of these are parts of a scenario, and the process for scenario development needs CAX support. Especially MEL/MIL development can be supported by a tool effectively. Moreover, the same tool can be used extensively for the management of an exercise because MEL/MIL constitutes the flow of an exercise and the work load of TA.

10.2.1 Exercise Management Tools

To achieve the exercise objectives, events and incidents are designed and injections are developed according to them before the exercise. "Events are major occurrences or a sequence of related incidents, which are situations that provide greater clarity to an event. An injection is the way of bringing an incident to the attention of players." Typically an event means several incidents and an incident is several injections. For example, an embargo can be an event for an exercise. Then, tracking and embarking to a suspected ship can become an incident for the embargo event. A detection report about a suspected ship heading toward the restricted zone is an injection for the embankment incident. All these injections, incidents, and events are listed in MEL/MIL that are usually not fixed, and they should be modified during an exercise (i.e., dynamically scripted throughout the exercise). A medium scale exercise is typically designed around 4–6 events, 50–70 incidents, and 1000 injections.

All the injections should be carefully synchronized with the simulation. In addition to this, many injections are created automatically by the simulation system based on the developing situation during a CAX. They should also be coordinated with the MEL/MIL drafted and maintained during the EP, the training objectives, and the workload on the TA. There are also other reasons to monitor carefully the injections created by the simulation system, which includes the following:

- EXCON staff needs to manage the incidents and events from the beginning to the end. It may be possible sometimes that an incident created from a developing situation by the simulation is missed at the beginning, and therefore it is not handled by TA and EXCON properly.
- Some incidents created automatically can hamper the exercise goals, and therefore they may need to be removed in advance before they come into the attention of the TA.

As it is clear from the previous paragraphs, MEL/MIL management is a key requirement that impacts not only the planning phase but also the execution phase of an EP. It is important also for the postexercise analysis and reporting phase carried out based on the actions taken by TA for the events and incidents. Therefore, a tool that can automate the MEL/MIL scripting provide the interfaces between the MEL/MIL scripts and the simulation, and collect the lessons identified from the execution of the scripts is important for CAX support.

The joint exercise management module (JEMM) of NATO fulfils almost all of these functionalities. It is widely used in NATO exercise and by many NATO nations. JEMM is a structured way to develop a well-balanced MEL/MIL that covers all the training objectives (TO) and TA fairly. First, TO and exercise objectives (EO) are entered to the system. Then every incidents has to be linked to a TO, which ensures that they are designed according to TOs. Moreover, every incident is designed in a structured way. There are leading injections and expected TA reactions (effect) for an incident as it is shown in Figure 10.2. If the desired effects are reached for an incident, the TA is released with rewarding injections. Otherwise, encouraging injections can be given. For example, a report about a violent crowd can be a leading injection for a crowd-control incident. If the TA takes proper actions, then a report about the dispersion of

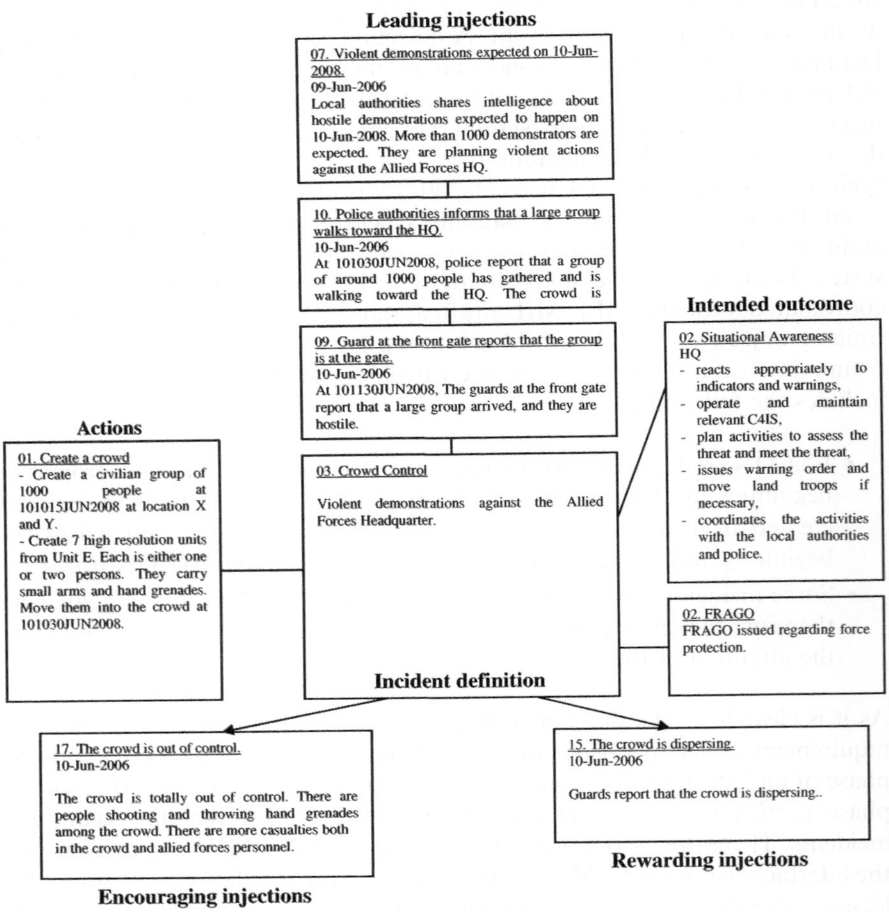

Figure 10.2: An incident in JEMM.

the crowd can be given. Otherwise, further violence can be reported as an encouraging injection.

Injections can be given by various means, such as media reports from media simulation (e.g., exercise TV channel, radio channel, newspaper, etc.), reports from lower level commands (LOCON) that represents subordinates or higher level commands (HICON) that represents superiors, situation forces (SITFOR) actions, or simulation system that automatically feeds C2 systems of the TA. The method to bring an incident to the attention of the TA is described in the actions part, especially when it needs to be synchronized with simulation. These actions are read by the response cells (RCs), and are taken as required. Note that injections that require the synchronization with simulation (i.e., actions) should not be given fixed dates/times and locations but stimulating conditions. Otherwise, the injection can be implemented unrealistically and synchronization of simulation requires magic move types superficial actions.

JEMM has also other supporting utilities, such as matrices that show the relation between the TO and incidents, intensity crated on TA at a certain time by a certain incident, and geographical locations of injections on a map. Another important module in JEMM is used by observer trainers (OTs) and event managers to accumulate observations about the effects. OTs are tasked to collect observations about incidents by using this module. They enter their observations to JEMM. These observations are analyzed by the event managers, and new injections either rewarding or encouraging are scripted dynamically if required.

JEMM is a very useful exercise management tool. It is not the only one. Other tools fall in this category, such as joint synchronized main event list (JSMEL) by U.S. Joint Warfighting Center (JWFC), an Exonaut by 4CStrategies of Sweden, and Mosbe by Breakaway of the United States.

10.2.2 Scenario Management Tools

The definition and the content of a scenario changes from a nation or an organization to another. A scenario typically is made up of some or all modules and submodules depicted in Table 10.3. Scenarios can be as follows:

- Real (i.e., based on real geography, nations and situations)
- Fictitious, often called generic
- Semifictitious (e.g., real geography but fictitious nations and events together with real nations and events)

In any case, developing a scenario that has the modules in Table 10.3 from scratch requires long time (e.g., 12 months), effort of a large team (e.g., 5–10 people), and some additional resources such as geodata, maps, and imagery. It also needs a careful management and coordination to maintain the consistency of the scenario.

A CAX support tool can shorten this process and ease many management and coordination challenges to keep the scenario consistent. NATO uses a tool

Table 10.3: Scenario modules.

Modules	Sub-modules
Geostrategic situation	Map Historical background Political, military, economic, humanitarian, and legal conditions
Theatre of operations	Map data set Theater data Country books Friendly and enemy order of battle
Strategic initiation module	Road to crises Related national or international documents and resolutions Strategic military assessment Initiating directive by the government/governing body Strategic planning guidance
Crises response planning information	Intelligence summary Friendly forces Civil/military, environmental assessments Reconnaissance reports/imagery Operational planning process (OPP) tool data sets
Force deployment information	Force list Deployment plan Intelligence summary Target list Rules of engagement authorization
Execution information	Road to war Intelligence summary Assessment reports Order of battle Situation reports Common operating pictures for C2 systems MEL/MIL data base

called joint exercise scenario tool (JEST) for this purpose. One very important contribution of JEST is the ability to define parameters. This list is dynamic, that is, there is no limitation for the number of parameters, where scenario developers some set parameters, such as D day, the names of nations, and so on. Then they use the names of parameters in the scenario modules. When they update the value for the parameter, it is updated automatically in all the modules where the related parameter is used. For example, critical dates, locations, and names can be defined as parameters. Later when it is required to change them, changing the parameter value suffices. This method saves time and ensures that the change is done everywhere required.

JEST also provides an effective way to generate scenario products from the fields in scenario. A developer can select the field from the available fields in the scenario data base and ask JEST to generate a country book in pdf format from those fields. Similarly, JEST can generate web pages for EXCON staff from the scenario data base.

JEST can also import order of battle (ORBAT) data from the data base of a simulation system. It is also possible to export ORBAT data edited by JEST to simulation, C2, and OP tools. This method helps the synchronization of many systems that need to work together consistently according to a common scenario.

The scenario tool must be in line with the EP because many scenario modules are related to each other and are prepared during different stages of the EP by different EPG members. For example, the geostrategic situation module should be an annex to an EXPEC. After the EXSPEC is issued, some other product parts of other modules are developed based on the geostrategic module. Intelligence reports are an example of this. An intelligent report must be consistent with the geostrategic module, and more than one intelligent report is produced as a part of crises response planning, force deployment, and execution information modules. The scenario development tool should fit in these requirements.

Scenario tools must also support the management of multiple scenarios, scenario modules that fit a given set of specifications, merging scenario modules from different scenarios, into a new scenario, and scrutinizing the resulting scenario for a new setting. These capabilities can enable rapid scenario generation, which is a critical and emerging requirement for mission rehearsal trainings (MRT) and mission rehearsal exercises (MRE). For rapid and expeditionary deployment of combined (multinational) forces requires MRE organized within days or weeks. The scenarios for such MREs must be prepared within several days. Note that a scenario is made not only of road to crises and country books but also of other modules like MEL/MIL.

CAX support tools may be useful not only for scenario management but also content development for scenarios. For example, simulation systems generate high-resolution and extremely realistic 3-D images. These simulation systems can be used to generate 3-D imagery as if they are taken by spy planes or unmanned aerial vehicles (UAVs).

JEST is not the only tool available for scenario management. Other tools like the terrain data management tool (TDMT) of the international solutions group (ISG) are available. Mosbe is not only a MEL-MIL but also a scenario tool. Other scenario generation tools that can generate scenarios automatically based on given specifications are also being developed. Please note again that the definition of scenario changes from one organization to another.

10.3 MEDIATION-WARE

SYNEX tools must replicate C4I environments during CAXs. In other words, simulation systems and all the other related software must be transparent to the TA. They must be able to carry out the exercise as they are in an operations and commanding their subordinates by using C4I systems normally available to them. The TA should receive orders and send their reports through C4I systems. This transparency can be achieved by the mediation tools between the simulation and C4I systems.

The interaction between constructive simulation systems and TA is depicted in Figure 10.3. As you can observe, there are two ways for this and both of them are indirect. The first way is through the response cells who are acting as LOCON to the TA. TA gives an order, and a response cell role plays as the subordinate unit headquarter that receives the order. They, the planners in the response cell coordinate and plan the execution of the order, and after applying some doctrinal C2 delays, they pass the plans to the CAX operators who translate them to the simulation orders. CAX operators receive both periodical and mission reports from the simulation system. They translate these reports to the planners in the response cell, who role play to pass the reports to the TA. This interaction with the simulation system is through the role play by the response cell.

The second way is automatic interaction through the C2 systems. Simulation systems can interact automatically with the C2 systems by using mediation ware. For example, the air tasking orders developed in a C2 system can be translated into the simulation orders and directly fed into the simulation. This

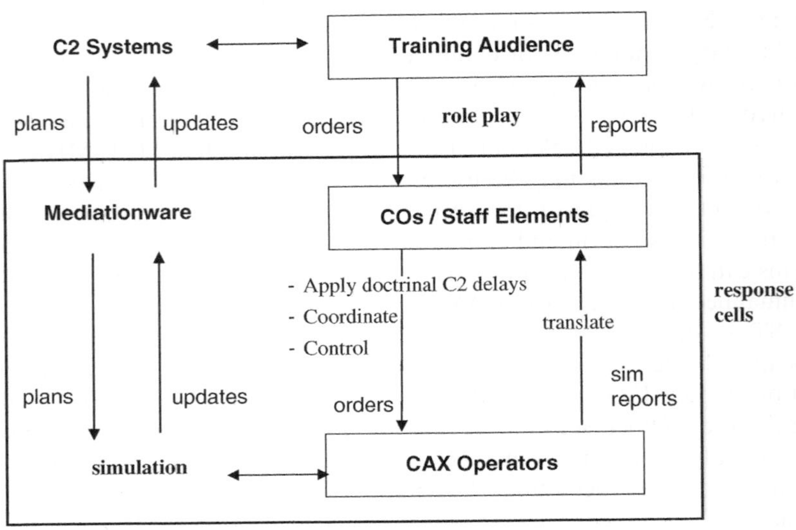

Figure 10.3: Simulation and the training audience.

interaction is from the C2 system to the simulation. The opposite direction is also possible. For example, the simulated air missions can update the recognized air picture automatically and continuously.

10.4 REVIEW QUESTIONS

10.1 Categorize CAX support tools, and explain each category. Give also the name of an example tool for each category.

10.2 Clasify constructive simulations. Do we still need categorizations as highly aggregated and high resolution? Discuss.

10.3 Is a federation of service models or a joint model more advantageous?

10.4 What is the most common function that exercise planning and management tools carry?

10.5 When the opposing sides of an exercise are represented by the training audience (i.e., opposition force is not a part of EXCON), is a MEL/MIL management tool like JEMM still useful?

10.6 Why is mediation ware between C2 and simulation tools are important? What can be done when a mediation ware is not available?

10.7 How can the results of simulation be reflected to the TA?

11

COMMUNICATIONS/INFORMATION SYSTEM ISSUES, TECHNICAL RISKS, AND RISK MITIGATION

Computer-assisted exercise (CAX) support involves many tools explained in Chapter 10. Each of these CAX support tools is more than a single software module. When federations are used, the software architecture becomes more complex. A medium-scale North Atlantic Treaty Organization (NATO) Training Federation (NTF) exercise where simulation systems are distributed in two sites, with more than 30 server software modules (i.e., programs), are run over 20 server machines located in remote locations. All these software modules have to work together collaboratively and coherently. They exchange data, and they are dependent on each other. All these make CAX support a complex and challenging task with many technical risks. One very important prerequisite to tackle the challenges in this environment is to have reliable and fault tolerant local and wide area computer networks that satisfy certain quality of service (QoS) metrics that include not only bandwidth but also end-to-end delay. In this chapter, we elaborate the technical challenges for CAX, technical risks, and some risk-mitigation techniques.

11.1 HARDWARE AND SOFTWARE REQUIREMENTS

Most of the constructive simulation systems and many federations can be run on a typical laptop computer. However, it does not mean that a single server machine is enough to run an exercise. A careful design of the hardware and software environment makes an impact on the performance of the CAX support tools. In this design, the key is to find the bottlenecks and to improve the performance on those points.

A typical constructive simulation system is made up of software modules that can be grouped as follows:

- Actual simulation server
- Data distribution server
- User (i.e., operator or controller) interfaces
- Mediationware

User interfaces are one for each user or controller. Therefore, they are run on a separate machine for each user, and they seldom become a bottleneck in an exercise. Most of the time, the bottleneck is either the actual simulation server or the data distribution server. Because the simulation server is typically a single module, it requires only one machine. Therefore, the only way to improve its performance is to increase the capacity of the hosting hardware. For the data distribution server, the story is different. A tiered architecture can be applied for this group: a primary tier and a secondary tier. The user interfaces can be grouped into clusters based on their locations, their functions, or both. Then, each cluster is linked to a data distribution module from the secondary tier. All the secondary tier data distribution software may then be linked to the primary tier. The primary tier is usually hosted by a single machine linked directly to the simulation server. This is illustrated in Figure 11.1. Note that the architecture for constructive simulations is not always like this. There are simulation systems where a single software is both a simulation and data-distribution server. There are also data distribution servers that do not allow a tiered design.

Many constructive simulation systems known to us run primarily on the Linux operating system. A typical programming language used for simulation and data distribution servers is C or C++. There are also simulation systems implemented by using programming languages like Simscript. User-interface software is implemented very commonly by using Java. If a data base management system is used for data base preparation and management, Oracle is preferred most of the time.

11.2 COMMUNICATIONS AND QoS REQUIREMENTS

When designing a network topology, the following should be considered:

- Cluster the network as depicted in Figure 11.1. When clustering is used, the primary data distribution server makes the updates only once for all the user interfaces linked to the secondary server. This technique reduces

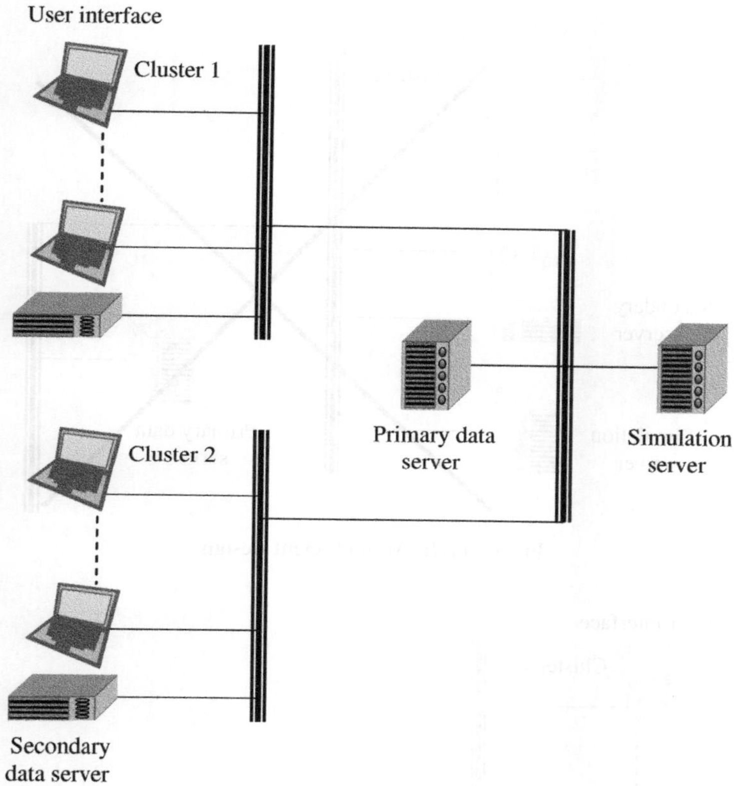

Figure 11.1: Hardware for a constructive simulation system.

the network traffic and is especially important when the data distribution is performed through a wide area network. If the tiered approach is not applied, then all the user interfaces download directly from the primary, which means one separate download for every user interface.

- Keep the servers that exchange data with each other in the same segment or network. For example, keeping the secondary data distribution and simulation servers in the same segment while the primary data distribution server is in another does not make much sense. The situations like the one in Figure 11.2 should be avoided. Figure 11.1 is a better design.
- Decrease contention on the links. This can be achieved by using multiple network interface cards (NICs). For example, the primary data server in Figure 11.3 is equipped with three NICs, i.e., one for each cluster and one for the simulation server. The simulation server is directly linked to the primary data distribution server by using a cross cable. Therefore, the data traffic for each cluster and server are forwarded to different segments, and they do not content for the same network resources (i.e., link) if switches

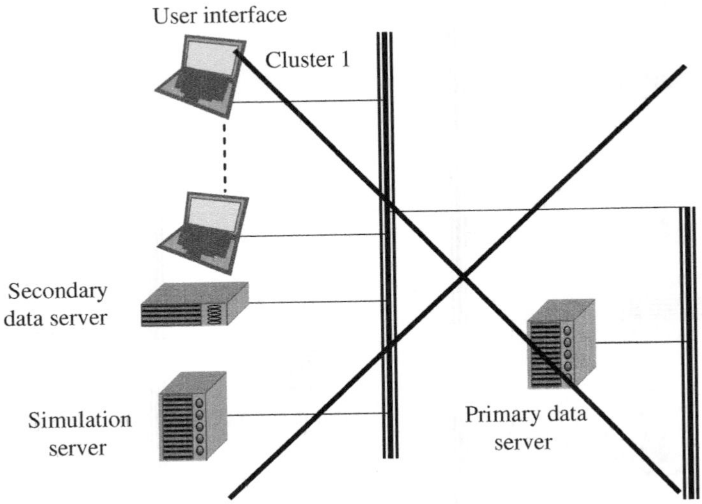

Figure 11.2: An inefficient design.

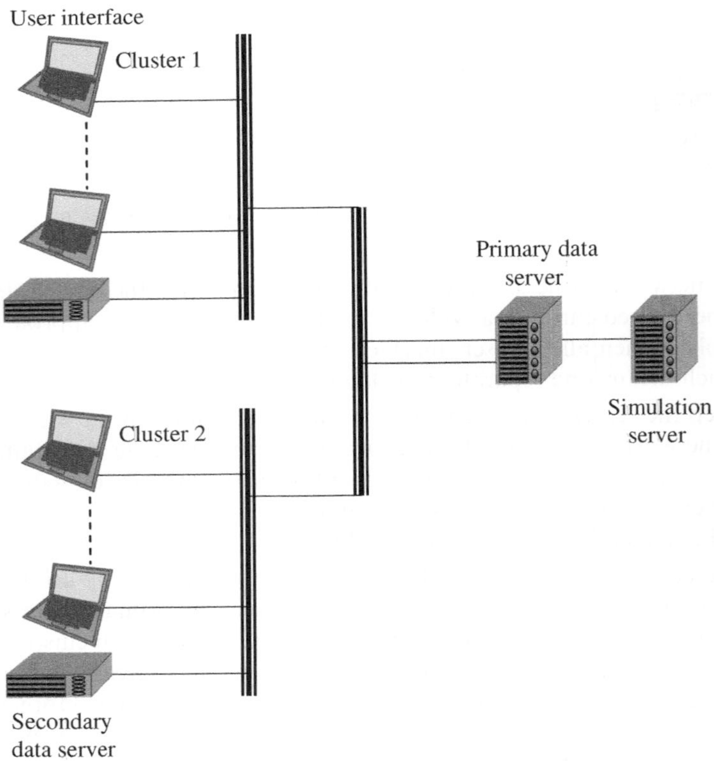

Figure 11.3: An efficient design.

are used, and NICs and routing tables are properly configured. This design is more efficient than the ones in Figures 11.1 and 11.2.

- Avoid designing the network based on the average data transmission requirements because the data traffic generated by constructive simulation systems are generally very bursty. When the game first starts, and at some certain times, all simulation data are transferred from the simulation server to the data distribution servers, and then the user interfaces. At this time, the available capacity (i.e., bandwidth) is used at the maximum level available. The larger the capacity is, the shorter the game start-up becomes. Therefore, the constraint for the time to start-up the game generally makes the bandwidth requirement, which is typically much larger than the average. Most of the time, the utilization of the available bandwidth is low because of this reason.
- Consider all types of QoS requirements. Apart from bandwidth, there are other QoS requirements. Especially the end-to-end delay constraint should be considered carefully because it may have an impact on the performance of the system and usage of the network resources as shown in Example 11.1. As you can observe, in the example as the delay increases the usage decreases. For the example, it is only 5%.
- Locate software modules according to the topology and link data. Two software modules that exchange data and can create low usage and poor performance should not be separated by a wide area network unless it is necessary.

Example 11.1

Two simulations are federated. Simulation **x** registers all its entities one by one and waits for an acknowledgment after sending each register message. The average time delay to receive an acknowledgement after completing the transmission of the register message is 2 seconds. The link capacity is 10 Mbps. The frame length for registration messages is 1000 bytes. Each registration message fits to a single frame.

a. If Simulation **x** has 500 entities to register, how long does it take to complete the registration of all entities? Assume that as soon as the acknowledgment is received, simulation x can start to transmit the next frame immediately without any delay.

Transmission of a single frame takes

$$t = 1000 \text{ bytes}/10{,}000 \text{ bps} = 0.1 \text{ second}$$

The total time required for a single entity is

$$a = 2 + 0.1 = 2.1 \text{ seconds}$$

The total time required for all entities is

$$T = 2.1 \times 500 = 1050 \text{ seconds} = 17.5 \text{ minutes}$$

b. What is the usage of the link during registration process. Ignore the communications and networking overheads:

$$U = t/t + a = 0.1/2.1 = 0.05$$

The bottom line is that the network design can impact on the performance of constructive simulation systems. The requirements and characteristics should be carefully determined, and the network should be designed based on these data. Often, simple relocation of servers can change the overall performance drastically.

11.3 SECURITY ISSUES AND CHALLENGES

When designing a CAX architecture, during and after a CAX several security related factors can make the design and the execution a challenging task. These factors are as follows:

- Accreditation of software: The software used for CAX can be considered in three layers: operating systems (OSs), ancillary software, such as data base management systems (DBMSs), and CAX support tools. OS and ancillary software are generally more stable (i.e., the new version releases are less in the number comparing to the CAX support tools). Typically, after every major exercise, there are some enhancements or bug fixes in the CAX support tools. If the software accreditation process cannot support the frequency of new software releases, then two things can be done:
  - The frequency of the software releases can be reduced.
  - The latest version of the software is not used. Certain number of versions skipped and software accreditation process can be started at a fix date every year and accredited software can be used after the process ends.
- Accreditation of hardware: Hardware should be selected according to the TEMPEST regulations. If the security regulations require, hardware may be formatted and degaussed after exercises.
- Network security issues: This part is perhaps the most challenging especially in international exercises, where not all the exercise participants are authorized to access all the resources and the architecture has networks with various classifications. There can be functional area services, such as MEL/MIL management tool that needs to be located at a certain network, and all the potential users may not have the clearance to access that

network. Sometimes, the availabilities and procedures may require connecting two networks with different classification levels, which may restrict the file transfers only to one direction. Typically, file transfer from a network to another network that has a lower classification level is not permitted. It may also be required to connect two networks that has totally different classification procedures and accredited by different organizations. Therefore, the availabilities of the network capacities and their security implications should be analyzed carefully for the design of a CAX architecture.

- Model classification: Constructive simulation systems may also include classified models, which means they may be classified. An exercise that has participants with a lower clearance level cannot be supported by those simulations.
- Classification of data in the data base: A CAX data base is does not consist only of order of battle (ORBAT) data. Therefore, it does not essentially mean that a data base that has only generic forces is an unclassified data base. There can be classified data in the data base although, ORBAT is generic. For example, the default aircraft loads, the default distribution of a force in a battlefield, the threshold data to change posture (i.e., the unit power level to change the posture from defense to withdraw), can be classified, which makes the CAX data base classified irrespective of the ORBAT data classification. The classification level of the data base must be coordinated carefully to prevent exposing classified data to the unauthorized audience.
- Classification of scenario: Exercise scenarios are more than CAX data bases. For example, the selection of geography, names, dates, the flow of events, and incidents can imply intentions, perceptions, and plans. Therefore, scenarios always have security implications.
- Classification of the doctrines and procedures: Even when everything including models, scenarios, and data bases are not classified, the plans can expose classified information about concepts and doctrines. Similarly, the procedures used in headquarters or information flow between the echelons may be perceived classified.
- TA clearances: Because of all the reasons explained in this section, the security clearances of the TA have important implications on the design of CAX architecture and organization, the work load, and time required to set the CAX environment. Note that the security implications of the two training audiences (TAS) that have the same level of clearance may still be different. A TA from a nation that does not have any clearance but is perceived as friendly can have different security implications from another TA from a nation that is a potential adversary. A need-to-know concept may also mean that the parts of TA from the same nation and that have the same level of security clearance are treated differently.

11.4 GAME CRASHES, CHECKPOINTS, AND CRASH RECOVERY

Any CAX tool can crash for many reasons, such as the following:

- Bugs in software
- Errors in data bases
- Wrong operator commands

Crashes in mediationware, CAX planning, and management tools are easier to tackle compared with simulation software crashes. When simulations crash during a CAX, the following should be done:

- The reason for the crash should be found, and its repetition should be prevented. Typically, a simulation crashes because of the last user command. Therefore, it is relatively easy to locate the piece of software code that makes the game crash. When the problem in the code is located, it may be fixed in a short time. If this is possible, then the software is recompiled and reloaded after the fix, and the game is restarted. If that is not possible, then the operators are warned about the command that causes the crash are and asked not to repeat the same command.
- The game should be replayed quickly until the point it crashed from the last available point where the game can be restored. During the game, many constructive simulation systems can save the status of the game together with the orders entered since the last save point. This point where the status of the game and orders are saved is called as checkpoint. It is generally possible to restore a game from a checkpoint. For many simulation system it is also possible to run the orders pending at the time when the checkpoint is taken. Since the same random numbers are generated unless the seeds are changed, the simulation generates the same results when it is fast forwarded. Some simulation systems especially federations may not have the capability to save and run the pending orders from the checkpoints. In that case, the orders entered after the last checkpoint should be reentered. If the reason that causes the crash is not fixed in the software, it is also required to remove the order that causes the crash from the order file before rerunning the game.

The timings of checkpoints are very important for recovery from a crash. They may also become very useful for analyzing the results later. There are two types of checkpoints, as follows:

- Periodic checkpoints. These are taken periodically. They can be as often as once in every 30 minutes or once a day. The frequency of the periodical checkpoints depends on several factors like disk space available and the time required to fast forward the game. If the disk space is a limiting factor, except for the last two three checkpoints the others can be deleted.

Apart from the last two to three checkpoints, the checkpoints required for data analysis (e.g., one for every day) can also be kept. Please note that it may not be possible to restore the game from the last checkpoint. Therefore, keeping at least last two to three checkpoints instead of the last one is recommended.

- Aperiodic checkpoints. These checkpoints are taken when important events that require many commands to the simulation are about to happen or when time-consuming computations are completed. The probability of a game crash increases when more than the typical number of orders are scheduled. Therefore, it is wiser to take a checkpoint just before the simulation starts to execute this burst of orders. For example, many air missions start almost at the same time. Before this, a checkpoint may be taken. If a checkpoint is taken when time-consuming computations are completed, it may become faster to fast forward the game if a crash happens after that point.

The bottom line is that a well-planned checkpoint strategy may help much in recovering from a crash.

11.5 SHADOW/RUN AHEAD GAMES

Another important risk-mitigation technique is the shadow/run ahead game. A shadow game can be run ahead of the actual game to determine whether the game crashes when executing the orders. Here, synchronizing the orders in the shadow game with the orders in the actual game is a challenge. This stage typically requires installing the checkpoints in the actual game and returning the shadow game to the checkpoint and restarting it from that point every time a new checkpoint is taken. Shadow games are extremely useful especially to check whether the orders in an air tasking order (ATO) may cause a crash. Every time an ATO is received, all the orders in the ATO can be entered also to the shadow game, and the shadow game can be fast-forwarded.

Shadow/run ahead games can also become very useful when certain shifts in the exercise are skipped. For example, except for the preplanned orders, nothing may be played during the night shift. In this case, response cells may leave in the evening and come back in the morning and start the exercise from the morning. The situation when they arrive in the morning may be different from the situation they left in the evening because of the orders entered previous day but executed during night. It may be very useful to brief the response cells and the SITCEN about the situation in the morning before they leave in the evening because the situation in the morning may require preparations. To do this when all the orders for the night are entered to the simulation, the run ahead game can be fast-forwarded until the morning, and SITCEN and the response cells can be briefed about the morning situation before they leave.

11.6 BACKUPS AND ARCHIVES

CAXs provide a huge amount of data stored in the checkpoints. These data can later be used to analyze the results of the exercise. These data can also be very useful to test and analyze other systems, procedures, or concepts. The following results are possible with the data collected by a simulation during a CAX:

- Generation of realistic patterns: Realistic patterns, such as unit mobility, equipment/combat system/unit availability, unit posture, and supply status/consumption pattern can be generated from constructive simulation data collected during an exercise. These patterns can be used to test new concepts and systems realistically, such as a new communications system or logistics concept [CAY02].
- Course of action/doctrine/concept analysis: Parts or whole of a CAX can be run on a simulation by changing some orders based on different concepts, courses of action, or doctrines. The impacts of these changes can be examined by analyzing the changes in the data related to the selected performance metrics.
- Simulation-based acquisition: Similarly, the impact of the changes in the capability of a combat system can be analyzed by using data from a previous CAX. For example, it is possible to analyze how changing the range of an attack helicopter can affect the overall results [DAG01].

Therefore, keeping the backups of the exercise data even long after the exercises may be very useful. The hint here is that it is also essential to keep the backup of the complete CAX environment for this purpose because constructive simulations are often not backward compatible for the data. Often with the new release of a constructive simulation, new fields are added to data base or some are removed. If the backup of the complete simulation is not available, it may need hard work to adapt the data to the new versions of the simulation. It is always wise to keep the backup of every versions of the simulation software for at least several years.

11.7 NETWORKING SERVICE OUTAGES AND OTHER REASONS FOR FAILURE

The risks during a CAX are not only limited to the game crashes. There are always risks like network or power failures. Hardware failures are also possible. The procedures for them should be prepared and rehearsed before a CAX.

To reduce the risk of hardware failure especially the server processes must be run on fault-tolerant server architectures. It is also recommended to have backup hardware available (i.e., two or three machines configured similarly to the actual and shadow game configurations). There are two options to recover when an actual game server machine fails, as follows:

- Configure one backup server and replace the failed server: Configuration may take time.
- Replace the failed server with the counterpart server in the shadow game, and then configure the backup server for the shadow game: Since shadow game is already synchronized with the actual game, this is a quicker procedure.

Several factors influence the selection of one of these options: time required to configure a backup server, time required to replace the actual game with the shadow game, the available hardware, the available man power for system administration, and so on.

For network failures, the redundancy is the first risk-mitigation technique. However, redundancy cannot guarantee that there will be no network failures. Therefore, keeping complete server environment in the same location when possible and having backups for the active devices for that location may reduce the risks. In this configuration when the network failures happen, the remote sites can be informed about the results from the simulation by the other means as if their primary connection to the battlefield is lost. If servers need to be located in remote locations, the simulation services cannot be provided until the failure is resolved. Therefore, when possible the systems should be configured such that they continue working locally when there is a network failure.

This may also require the movement of the people during network failures. For example, let us assume that RCs and SITCEN are located in different buildings. There need to be plans for the case that the network between these buildings fail for an extended period of time. This may require collocating the representatives from the RCs and SITCEN in an environment where network services can be locally provided until recovery.

All network services and power may be lost for an extensive period of time. We call this a catastrophic failure. During a catastrophic failure, the minimum configuration required can be provided through portable servers and workstations. Their batteries can be recharged and/or replaced. Hence, the minimum services can continue at least for a reasonable time period.

11.8 REVIEW QUESTIONS

11.1 There are three remote sites in an exercise. In each remote site, there is a data repeater for all the workstations at that site. Data repeaters receive data from a primary data server. The simulation server sends the data only to the primary data server. Design a topology for this simulation network.

11.2 Discuss various quality of service parameters for data communications and their importance for live-virtual-constructive federations.

11.3 There are two types of training audiences in an exercise. One of them can see classified data restricted to the other. Discuss the networking challenges for such an exercise.

11.4 Classify checkpoints. Why are they needed?

11.5 When and why are shadow and run ahead games needed?

11.6 List the possible failures during a CAX, and explain principles for the design of recovery procedures for each of them.

12

EXERCISE CENTERS AND FACILITIES

Whenever planners receive an order to plan and organize a computer-assisted exercise (CAX) for a certain training audience, they should think first about proper physical conditions for the execution. It is of utmost importance that the exercise centers and all facilities provide holistic conditions for achievement of desired exercise and training objectives.

Training/exercise centers have been typically formed for conduction of a subset of the following functions:

- *Exercise planning*: Exercise centers can be designed to plan an exercise according to a given exercise specification. This includes preparation of important products, such as exercise plans (EXPLAN) and scenarios including master event list (MEL) and master incident list (MIL) as its extension.
- *Training*: Exercise centers can provide training support before, during, and after an exercise process, which may include the following:
    - Training/mentoring support, observation, and evaluation during the operational planning process
    - Training on the functional systems, such as a command and control system or a collaboration tool
    - Training on functional area, such as a headquarters (HQ) function
    - Training on the procedures how to interact with the other functional areas
    - Training of HQ for a specific mission

Computer Assisted Exercises and Training: A Reference Guide, by Erdal Çayirci and Dusan Marincic

- ***Exercise management***: Exercise control staff can be formed and/or managed by an exercise center.
- ***CAX, communications and information system (CIS), and real life support for exercises***: CAX, CIS, and real life support are the set of services that an exercise center is supposed to provide at a minimum. The support can be provided only for the exercise control staff or can be as large as the overall support for the exercise.
- ***After action review***: A training center can be tasked with analysis, evaluation, and assessment of the training audience, and/or manage a data base for lessons identified and learned.
- ***Experimentation***: Exercises provide good opportunities to experiment with new concepts, capabilities, and tools.

The set of tasks given to a training/exercise center is the most important factor that influences the organization of the center and the design of facilities for the center. This set of tasks is dependent on the hierarchical position of the center within command hierarchy. There are several approaches to locate a training/exercise center in an organization, as follows:

- **Subordinate to a command for transformation or a training/doctrine command**: If armed forces have a command responsible from transformation that develops new concepts and capabilities and provides training for the force structure, the command often has one or more training/exercise centers in various levels. Usually, these centers are tasked with all the functions listed above. There may be a separate center for each level of training and exercises (i.e., strategic, operational, and tactical) or a center can be tasked with all levels. If there is a separate center for each level, they may be colocated or not. In any case, good coordination and collaboration are essential for the effectiveness and success in accomplishing the training objectives.
- **Military academies, staff colleges, and branch schools**: Almost every military academy, staff college, and branch school of modern armed forces has either a modeling and simulation center or is supported by a training/exercise center. Sometimes the modeling and simulation centers in military academies and staff colleges are also tasked to provide modeling and simulation support for the exercises of other units and headquarters.
- **Civil training centers**: Apart from the centers for military and police, the number of civil training centers that use simulation and CAX technologies is also increasing. These centers have been focused mainly on crises response and consequence management operations. They are also on various levels, such as political, strategic, tactical, operational, and technical. Please note that civilians consider operational level as a lower level than tactical. See Figure 12.1 for a listing of military and civil training perspectives.

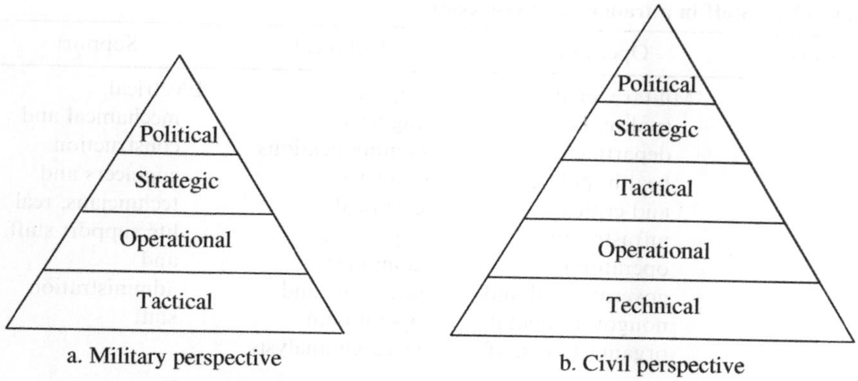

Figure 12.1: The levels for training centers.

The number of training/exercise or modeling and simulation centers is higher for the lower levels. There is typically one center for strategic level in a country. However, there may be many centers for technical level.

12.1 ORGANIZATION OF A TRAINING/EXERCISE CENTER

As few as five or six technical staff (i.e., engineers and technicians) can form a modeling and simulation center for a school or headquarters. In this case, the technical staff is only responsible to set one simulation system for exercises and manage it through exercises. When the center is a part of training/doctrine or transformation command and is tasked with all the functions explained above, a much larger and more complex organization with hundreds of personnel is required. These personnel can be categorized into three broad groups, which include operational, technical, and support. The command group of the organization should be from the operational background that represents the intended training audiences. Ideally the command group should have a real experience about the operations in the training scope, and the commander should be at least equal in the rank to the highest rank in the training audience. The commander becomes the directing officer for the exercises and takes an active role for the training of the highest echelon in an exercise. See Table 12.1.

12.1.1 Operational Staff in a Training/Exercise Center

Permanent staff members with an operational background constitute the core of a training center. They make a plan and manage the execution of the plan for an exercise. Their roles and responsibilities can be listed as follows:

- **Exercise planning, management, and reporting**: A team manages an exercise process from the start to the end. The team leader is the officer that has the

Table 12.1: Staff in a training/exercise center.

Category	Operational	Technical	Support
Background	Military, civil (police, fire department, health, politics, and critical infrastructure operators), governmental and nongovernmental organization staff	Computer engineers, communications engineers, electrical engineers, computer scientists, and operational research analysts	Electrical, mechanical and construction engineers and technicians, real life support staff, and administration staff
Functions	• Command • Mentors • Exercise planners • Scenario preparation • Trainers/observers • Lessons learned analysis • Experimentation	• CAX support • Communications and information systems (CIS) support • Information security • Scenario support (geology and meteorology) • Media simulation • Data analysis	• Exercise set up • Real life support • Physical security • Operations and maintenance • Public information • Visitors office

primary responsibility to the commander for successful management of the assigned exercise process. The team provides the coordination among all the parties involved in an exercise, plan and coordinate the resources including manning, prepares and maintains an EXPLAN, manages the execution of the EXPLAN, and prepares the final exercise report. Two factors influence the required number of exercise planning teams n_t in a training/exercise center: The length of an exercise process l_e [(l_e is in months for Equation (12.1)] and the number of exercises n_e that the center runs a year.

$$n_t = \frac{l_e}{12} \times n_e \vdots \qquad (12.1)$$

An exercise team is busy with the exercise from the start to the end. Therefore, they cannot be involved in more than one exercise process. For example, if the exercise process is 12 months, then a team can be involved in one exercise per year. Hence, if the center is conducting five exercises a year, there should be at least five teams in the center. Each team should have at least one team leader, one member for resource coordination and management, and another member for the management of the exercise plan (i.e., version control).

- **Scenario preparation including MEL/MIL**: Scenario preparation is another important task for the operational staff at a training/exercise center. Scenarios must change every time that the same headquarter/unit becomes the training audience. The scenario can be based on real geography and geostrategic situation or a fictitious setup. The fictitious scenarios are often called generic. In any case, preparing a good scenario from scratch takes a long time, generally more than a year. Moreover, scenarios must be modified for every exercise, and the staff responsible for each scenario must be available during an exercise to respond to requests for information (RFI) coming from the training audience or EXCON. The scenario team should include subject matter experts on topics, such as maps and geographic data; meteorology; intelligence; land; air; and naval operations; engineering and CIS; civil–military cooperation issues; police, health, and fire department issues; international governmental and nongovernmental organizations; political issues; and the media.

  MEL/MIL is also a part of the scenario, but it is not prepared by the scenario group. MEL/MIL represents an extension of the exercise scenario that should fulfill all the training objectives throughout the exercise and realistically activate the training audience operational procedures. Therefore, it is prepared by the training personnel in close coordination with both exercise planners and scenario staff. Since the MEL/MIL drives an exercise, CAX support personnel is also in close coordination with the training staff responsible for the creation of the MEL/MIL.
- **Training, observation, and evaluation**: The training team is responsible for the training methods explained above. They support the operational planning process and become a part of EXCON as observers/trainers during the execution phase of the exercises. Their structure should cover the entire training audience to assist them to achieve the desired training effects and each functional area.
- **Mentors**: Training/exercise centers should also have a pool of well-experienced senior officers who can be called to support the training of the commanders during the execution phase of an exercise.
- **Lessons identified and learned management**: Observers and trainers collect a high number of lessons identified during each exercise. These lessons should be carefully analyzed, and required actions should be taken accordingly. The management of a "good lessons learned" data base can be very useful to manage the other exercise processes more efficiently and become an important aid for the training audience to get prepared for an exercise.
- **Experimentation**: This staff designs experiments, conducts them, and collects the data; analyzes the collected data; and reports on them. The experiments can be carried out during an exercise or off-line by using data collected during an exercise. Therefore, the formation of the

experimentation team is dependent not only on the level of ambition for the exercises but also on the level of ambition for experimentation.

12.1.2 Technical Staff in a Training/Exercise Center

The role of technical staff in a training/exercise center is at least as important as the role of the operational staff, and their job is typically more challenging compared with an operational headquarters or unit because of the following reasons:

- All the functional area systems (FASs) available in all headquarters and units that may become a training audience should be replicated in the center. Therefore, the number of FASs in a training/exercise center is normally higher than the number for a headquarters or unit.
- In addition to all FASs, a training/exercise center should also be equipped with all sort of CAX support tools.
- Finally, none of the FAS or CAX tools are static. They need to be reconfigured for each exercise.

Technical staff in a training/exercise center can be classified as follows:

- **CAX support**: A well-skilled CAX support personnel should be a person with an operational background and should also be educated on operational research and computer science. This skill set is rare. The minimum skill set for a CAX support person can be a computer engineer trained also on operational research or vice versa. Of course, CAX support personnel should also be very well trained on the CAX support tools and on the ancillary software related to these tools. They prepare the data bases, configure the CAX systems for exercises, and run CAX support tools during the exercises. For each exercise, a team of minimum three engineers (i.e., a team leader, a database engineer, and a technician) that configures the CAX tools is required. Like on the planning team, a CAX support team can also concentrate only on one exercise at a time. Therefore, the number of CAX teams in an exercise center should ideally be equal to the number of planning teams. Apart from the CAX planning teams, the operators are also needed during the execution phase of an exercise. The number of operators is based on the number of response cells and their complexity. Two or three operators are needed in each response cell. For situation forces (SITFOR), a higher number of operators is needed. There are two approaches for providing CAX operators for an exercise, as follows:
    - They can be augmented and trained before the exercise. Because they are often not skillful enough, avoidable game crashes may occur, and CAX conduct may not be as efficient as it can be.
    - Operators can be professionals who work for the center. This approach is of course more costly but safer.

- **CIS support**: Information technology (IT) systems, communications, and networking become very challenging especially when an exercise is distributed. The challenge is even bigger when parts of the training audience come from different nations, and when they participate in the exercise by using their own command and control information systems (C2IS) and communications systems. A CIS planning team that collaborates with the exercise planners throughout the exercise process is the first requirement to tackle with the requirements of this complex environment. This can be a small team of two to three people. They can focus on a single exercise at a time the same as the exercise planning teams. Apart from the planners, a large group of people responsible for the implementation of the plans about the following issues is also needed:
  - Installation and configuration of the IT and communications equipment
  - Installation, configuration, and administration of FAS such as C2 tools, operational planning process (OPP) support tools, and so on
  - Installation, configuration and administration of local area networks
  - Leasing, installing, configuring, and administrating lines and active devices for wide area communications
- **Information security**: Information security becomes typically the first obstacle when introducing any new CIS hardware (HW) and software (SW), networks, and FAS. Both computer security and communications security are critical issues during exercises, and they become very difficult to tackle when an exercise is classified and distributed. The number of people in this responsibility depends on the accreditation procedures and responsibilities.
- **Scenario support including maps, geographic, and meteorological data**: Electronic maps, geographical and meteorological data generation, procurement, processing, and maintenance are another specialty area. These tasks require a large organization. Therefore, typically services for these tasks are received from third parties, and those services are managed by one or two experts in the center.
- **Media simulation**: Almost every modern training/exercise center simulates media. For this, one or several TV stations and one or several newspapers are simulated. Because available issues are reused after being scrutinized, and there is not time pressure for preparation, media simulation can be managed by a small group of well-equipped people.
- **Data collection and analysis**: A huge amount of data can be collected during an exercise by simulation and C2 systems. The analysis of these data by experts can be very useful for many purposes, such as follows:
  - Evaluation and assessment of training audience
  - Fine tuning of simulation databases

- Off-line experimentation, such as experimentation for new CIS equipment by using the mobility patterns generated from previous exercise data
- Performance analysis for equipment in the exercise data base and courses of action taken during exercise

12.1.3 Support Staff in a Training/Exercise Center

Exercise centers normally host all exercise control (EXCON) and a subset of training audience (i.e., all of it or a part of it) during exercises. This implies the requirement for a well-organized support system that can provide the following:

- **Exercise set up**: The exercise area should be set up according to exercise requirements, which includes moving furniture, marking the exercise areas and routes, and so on.
- **Real life support (RLS)**: RLS includes many challenging tasks, such as arranging accommodation, transportation, food, and laundry services.
- **Physical security**: All the exercises and especially the international exercises have security risks that may come from outside, inside from an exercise participant, and from inside to outside. Exercise participants should be protected against physical attacks that may come from outside or inside. And the risks created by the exercise against the environment and external entities should be analyzed and mitigated carefully.
- **Operations and maintenance for the facilities**: Warming/cooling, lightning, cleaning, and maintaining the facilities are also among the important tasks of support personnel.
- **Public information**: Public information can be divided into internal and external. Sometimes, public information becomes very time consuming for the command group when the exercise is broadly publicized, because of the timings, the configuration of training audience, and contemporary events. Then, the public information office can become one of the key role players in an exercise.
- **Visitors' office**: There are almost always visitors in exercises, and Distinguished Visitors Day (DV-Day) is the most stressful day for anyone attending the exercise (i.e., EXCON or TA). A good visitor's office that carefully organizes programs for visitors can prevent TA and EXCON from being distracted from the exercise objectives.

12.1.4 Number of Teams and Staff in a Training/Exercise Center

We can generalize Equation (12.1) that gives the required number of teams or personnel in a training/exercise center for any functionality, such as the scenario and/or MEL/MIL preparation as follows:

$$n_t = \frac{l_e + l_i}{12 - l_o} \times n_e \vdots \tag{12.2}$$

where

n_t is the required number of team or personnel
l_e is the length of involvement in months for each event
l_i is the length of time in months required for preparation between the events
l_o is the length of holidays and other time off in months for a team/person in a year
n_e is the number of events that the center runs in a year

Of course, Equation (12.2) gives the minimum number for the ideal case where the events are scheduled evenly. The same equation can also be used to derive another equation for the maximum number of events that a center can host a year:

$$n_e = \frac{12 - l_o}{l_e + l_i} \times n_t : \tag{12.3}$$

where

n_e is the number of events that the center can run in a year
l_e is the average event length in months
l_i is the length of time in months required for preparation between two events
l_o is the length of time off in months for purposes, such as, facility maintenance
n_t is the number of events that can be run simultaneously

Example 12.1

An exercise center can host two events simultenously. The average event length is 2.5 months. For each event, at least 15 days are needed for build-up and tear down. August is a block holiday month for the facility.

a. How many events a year can be hosted in this facility at a maximum?

$$n_e = \frac{12 - 1}{2.5 + 0.5} \times 2 = 7 \text{ events}$$

b. Each event is planned and executed by an event planning team. An event process takes 5 months, and one event planning team can work on a single event at a time. After an event process is completed, the event planning team needs at least 15 days before starting with the other event. August is a block holiday, and all the event teams are in holiday in August. How many event planning teams are needed to use the exercise training center 100%?

$$n_t = \frac{5 + 0.5}{12 - 1} \times 7 = 4 \text{ event planning teams}$$

12.2 DESIGN PRINCIPLES FOR TRAINING/EXERCISE CENTER FACILITIES

Facilities are also very important for the overall success of the exercise. One may argue that there is no need for special facilities for an exercise center because moving the training audience costs a lot; therefore, exercises should be conducted where the training audience is. If there are multiple training audiences in an exercise, distributed simulation and exercise techniques can be used, and the exercise can still be conducted without moving the training audience. This is a viable argument and it prevents accommodation and travel costs incurred because of moving the training audience to an exercise center. However, a good training/exercise center still needs training/exercise facilities for the following reasons:

- The staff of training/exercise center needs offices.
- Exercise equipment needs to be stored and prepared in a facility.
- When exercises are conducted in a central facility, a smaller organization can suffice for the center, and travel/CIS costs of the exercise center are reduced considerably.
- Exercise centers are often better for information security.
- EXCON staff needs to be gathered in a facility and to be provided with EXCON tools.
- Rehearsing the deployment procedures can be an exercise objective for the training audience.
- Training audience can be better immersed in exercise when they are away from their normal locations.
- Finally, the training audience can be a composite headquarter formed for a specific mission.

Training/exercise center facilities must be large enough and equipped well enough to host their largest potential training audiences and EXCONs. The most important design principles for a training/exercise center can be listed as follows:

- **Flexibility**: Flexibility is essential, because almost every time the training audience changes, a new set up of the training environment is needed. The designated exercise space organized as an air operations center for one exercise can be reorganized according to the operational setup of headquarters for another exercise. Moreover, as time passes, requirements and equipments change. The center should be easily adaptable for the changes in requirements. This can be achieved by the following:

- Instead of small rooms, large halls divided by movable separators or walls should be preferred.
- All furniture must be portable. They should be easily assembled and disassembled and must be robust enough for reassembling and moving around many times.
- All equipment, power sources, hardware, and software should be easy to expand. Everything should be expandable and shrinkable.
- All communications, power, and networking lines should be laid in accessible shafts. Laying a new line should be very easy and without destructing anything.
- All kinds of networks in the scope of all potential exercises should be either available or easily connectable. For example, if there is an operational network needed for an exercise, and it is not available in the center, connecting that network to the center should be as easy as patching, installing few active devices, and accrediting. Such a requirement should not incur a need to install a new fiber line between the center and a telecom organization.
- Seminar halls must be large, divisible, and reconfigurable.
- The auditorium must be large enough for the largest expected training audience and EXCON or perhaps larger than that.

- **Immersion**: Everything should be designed in such a way that the training audience is not distracted from the exercise situation anytime. Distraction may happen when a distinguished visitor is briefed; an observer/trainer gives instructions or visitors wander around. All these can be isolated by providing seminar halls and syndicate rooms for briefing visitors and by giving instructions to parts of the training audience without disturbing the entire training audience. Everything in the exercise halls should be operation related.
- **Maintainability**: Designs must be simplistic and robust. Every equipment, line, and hardware should be marked and documented carefully. The facilities should fit the level of ambition for the center and be maintainable by the center. To be on the safe side, there is a tendency to procure more capacity than required. This approach is difficult to maintain, and a lack of discipline in the usage of equipment hinders the efficiency in using personnel. Therefore, everything being procured should ensure conduct of events within no later than 2 years.

The capability of a training/exercise facility is more dependent on the systems available in the facility than on the facility itself. Important systems for a training/exercise center include the following:

- **Communications**: A state-of-the-art training/exercise center must have either all the communications capabilities or the infrastructure for those capabilities being used or will be used by the potential training audience. This includes the Internet, all sorts of intranets available to the training audience, line of sight, satellite and high-frequency (HF), very high-frequency (VHF) communications. Analog or digital telephones, voice over IP, teleconference, video teleconference capabilities, and expandable large infrastructure to support all of these are required. When security regulations allow, low tier mobile communication systems like digital enhanced cordless telecommunications (DECT) or a similar technology [e.g., personal access telecommunication system (PACS), personal handy phone system (PHS), etc.] is also recommended.

- **Local Area Networks**: A set of efficient broadband local area networks is perhaps the most important technical requirement in a training/exercise center. There should be more than a single network because when an exercise is being conducted, another one may be prepared. The correct number of networks is given by Equation (12.4):

$$n_n = \frac{l_e + l_p + l_t}{12 - l_o} \times n_e : \tag{12.4}$$

where

n_n is the required number of network
l_e is the average length in months of events
l_p is the average length in months required for preparing an event
l_t is the average length in months required for tear down
l_o is the length of holidays and other time off in months for the center
n_e is the number of events that the center runs in a year

These networks can be physically decoupled, and the classification level may change from one exercise to the other. Efficient procedures and degaussing tools have to be in place to wipe out a network before making it available for an exercise, with a lower classification level.

When security regulations allow, wireless networks [CAY09], such as WiFi (IEEE 802.11), Bluetooth (IEEE 802.15.1), ZigBee (IEEE 802.15.4), and WiMax (IEEE 802.16), may become very useful. Interested users can refer to [Cay09] for the security risks in these kinds of networks.

- **Functional area systems**: It is always a good idea to replicate all FAS available to the training audience also in training/exercise centers. They should look like exactly as they are in operations. During exercises, the scenario is uploaded into them. If the training audience is not deployed into the training center, they use their operational FAS initialized by the exercise scenario in the center. Still, exercise centers need a replication of FAS to prepare and test the exercise databases on them.

- **CAX systems**: All kinds of CAX tools explained in Chapter 10 should be available in appropriate levels in a training/exercise center.
- **Virtual systems**: For virtual tools, not only hardware and software but also special space and facilities are required. When training/exercise facilities are being designed, special requirements for the virtual systems should also be considered.
- **Data security**: All kinds of data security systems, such as the following, should be available in a center:
  - Degausers or special software to wipe out the hardware after events
  - Scanners to scan incoming and outgoing storage media and computers against illegal data transfers, viruses, and trojans
  - Devices to monitor electromagnetic emissions from the center
  - If necessary, jammers and electromagnetic emission detectors

The design of a building can also help information security. Highly classified data can be located more central locations (i.e., in the middle of the facility). The rooms and halls, where classified data are stored, processed, or talked about, should not have windows. All TEMPEST regulations should be applied.

- **Physical security**: For physical security a well-designed access control system is required. This system can be based on cards, radio frequency identifications (RFIDs) or biometrics. It must allow the exercise managers to define who can enter which facilities, and even to follow the whereabouts of exercise participants. The security system should not create bottlenecks when many people try to access the building. They should be supported also by well-designed procedures to in-process incoming exercise participants smoothly and quickly. Facilities should also be designed accordingly. Facility approaches, doors, and entrance halls should be arranged such that incoming personnel are classified according to their functionalities and are not mixed. For example, if there are multiple sides playing a war game, each side can enter the building without mixing with the other sides and EXCON. Physical security measurements should not hinder the evacuation in case of emergency.
- Security systems should also be bidirectional, which means that exercise participants can be scanned both when entering and exiting the facility. Camera systems and perimeter intruder detection systems can be used to monitor more accurately all the access routes to the facility. Camera systems are also needed inside the building. Fire detection and extinguishing systems should be designed according to data safety needs. In locations where valuable data are stored, dry fire extinguishing systems should be preferred.
- **Power systems**: Fault-tolerant power systems, including uninterruptable power supplies (UPS) and generators, are also another important requirement. For Tempest, power line filters should be used to decouple the power lines for the computers with a security classification from the power lines of computers connected to the Internet.

- **Cooling systems**: A well-segmented cooling system is a requirement. When cooling systems are designed, it should be considered that any room can be filled with many IT systems. If every room is cooled equally, whereas one of the rooms packed with much electronic equipment is too warm, the others may become too cold.
- **Auditorium and seminar rooms**: Auditorium should be equipped with systems like simultaneous interpretation, projection, lightning, and sound and stage systems. Because auditoriums in exercise facilities are large and not used 100% for exercise purposes, it is a good idea to design them as a separate block with foyer and other facilities so that it can be used for conventions, seminars, and so on. This also requires a separate entrance system to the auditorium.
- **Park**: Park is also a critical part of an exercise training center. It must be a large well-leveled area that allows large deployable command posts can be easily deployed. There must be data and power plugs (i.e., wall and floor sets) where a deployable module can plug into the network in the facility. In the vicinity, there must be also space to locate satellite dishes.

12.3 REVIEW QUESTIONS

12.1 List the functions of a training/exercise center.

12.2 What kind of training tasks can be given to a training/exercise center?

12.3 What can be the position of a training/exercise center within command hierarchy?

12.4 Is operational or tactical higher echelon? What is technical level?

12.5 Do operational or technical people have a more important role in a training/exercise center? Who should be part of longer term staff? Why?

12.6 What are the roles of CAX support personnel?

12.7 What are the roles of support staff?

12.8 In an exercise center, only one event can be conducted at a time. Each event takes 1.5 months. For each event, 2-week build-up and 1-week tear-down periods are required. The center is functional all year. What is the maximum number of events that can be conducted in this center?

12.9 A CAX planning team can be involved only with one exercise at a time. After the CAX planning team is involved, the exercise process continues for another 5 months. After each exercise process, a CAX planning team needs another 15 days for reporting. Personnel have a 1-month leave every year. What is the minimum number of CAX teams required to achieve the maximum number of events calculated for Question 12.8?

12.10 Explain and discuss the design principles for a training/exercise center facility.

REFERENCES

[ALB01] Alberts, D., and Gartska, J. J. "Network Centric Warfare:Report to Congress," Washington, DC: Department of Defense, 2001, pp. 3–7–3–9.

[ALB02] Alberts, S. D., and Hayes, R. E., "Code of Best Practice for Experimentation," Washington, DC: CCRP, 2002, p. 61.

[ALB05] Alberts, S. D., and Hayes, R. E., "Campaigns of Experimentation," Washington, DC: CCRP, 2005, p. 2.

[BEN78] Bennett, R. J., and Chorley, R. J., "Environmental Systems. Phylosophy, Analysis and Control," London: Methuen & Co, 1978, p. 251.

[BIR98] Birkel, A. P., "SEDRIS Data Coding System (SDCS)," The MITRE Corporation, 1998, pp. 1, 2, 9.

[BOO73] Boorman, S. A., and Oliver, D. C., "Metrics on spaces of finite trees", J. Math, Psych., Vol. 10, Oct. 1973.

[BUS08] Buss A., and Jackson L., "Distributed simulation modeling: A comparison of HLA, CORBA and RMI," Proceedings of Winter Simulation Conference, 1998.

[CAR92] Carley, M., and Christie I., "Managing Sustainable Development," London: Earthscan Publications Ltd., 1992, pp. 129, 135.

[CAR04] Cares, R. J., "An Information Age Combat Model," Washington, DC: Alidade Incorporated, 2004, pp. 2–5.

[CAY02] Cayirci, E., and Ersoy C., "Simulation of tactical communications systems by inferring detailed data from the joint theater level computer aided exercises," *SCS Simulation J*, Vol. 78, 475–484, (2002).

[CAY07] Cayirci, E., "Multi-resolution exercise control structure for NATO education and training network." In Improving M&S Interoperability, Reuse and Efficiency in

Support of Current and Future Forces (pp. 12-1–12-18). Neuilly sur-Seine: RTO, 2007.

[CAY09] Cayirci, E., and Rong, C., "Security in wireless ad hoc and sensor networks," New York: Wiley & Sons, 2009.

[CHO71] Chorley, R. J., "Physical Geography. A Systems Approach," London: Prentice-Hall International Inc, 1971, p. 370.

[COB98] Cobb, L., A Scale for Measuring Very Rare Events, Couisuille: Etheling Consultents.

[CRA02] Crawshaw, J. and Chanbers, J. "A Concise Course in Advanced Level Statistics," Nelson Thornes, UK, 2002.

[CRI01] Crichton, M., "Training for decision making during emergencies," Aberdeen: University of Aberdeen, Department of Psychology, 2001, p. 8.

[DAG01] Dag L., and Cayirci E., "Performance evaluation of attack hellicopters in joint theaters by using constructive simulation," Realistic Rotorcraft Simulation Conference, 2001.

[DAV89] Davies, R., and O'Keefe R., "Simulation modelling with Pascal," Hertfordshire: Prentice Hall International Ltd., 1989, p. 1.

[DYK81] Dyke, B., "Computer simulation in anthropology," Annual, 1981, p. 193–207.

[DOC06] Doctrine, Development, Education and Training Command, "Military Doctrine," Ljubljana: Defensor, 2006, pp. 9–10.

[DOY00] Doyle, M. W., and Sambanis N., "International Peace Building: A Theoretical and Quantitative Analysis," Princeton in Washington, DC: American Political Science Association, Princeton University, The World Bank, 2000, pp. 9, 45.

[EDM88] Edmonds, M., "Armed Services and Society," Leicester: Leicester University Press, 1988, p. 122.

[ENL90] Enloe, C., "Policija, Vojska i Etnicitet (Police, Military and Ethnicity)," Zagreb: Globus. Nacionalna i sveučilišna biblioteka (National University Library), 1990, p. 147.

[ETA05] Etat Major de L' Armèe de Terre, French Concept Of Land Forces Employment In The Stabilization Phase, Paris: Ministere De La Defense, 2005, p. 8.

[GAL73] Galbraith, J. K., "Economics and The Public Purpose," Boston: Houghton Mifflin Company, 1973, p. 15.

[GAS96] Gass, S. I., and Harris, C. M., "Encyclopedia of Operational Research and Management Science," Norwell: Kluwer Academic Publishers, 1996.

[GEO67] George, F. H., "The Use of Models in Science. Models in Geography," London: Methuen & Co., 1967, p. 43.

[GRI99] Grizold, A., "Evropska Varnost (European Security)," Ljubljana: Faculty for Social Science, 1999, p. 2.

[HAG67] Haggett, P. and Chorley, R. J., "Models, Paradigms and the New Geography. Models in Geography," London: Methuen & Co., 1967, p. 19.

[HAM07] Hammes, T., "Fourth Generation Warfare Evolves, Fifth Emerges," Military Review, 2007, pp. 14–23.

[HOW94] Howard, M., and Guilmartin, J. F. Jr. "Two Historians in Technology and War," Carlisle: Strategic Studies Institute, 1994, p. 1, 2.

[IBI02] Ibid, Part two—Strategy 2020: Canadian Defense into the 21st Century, p. 9.

[IEE95] IEEE Ltd 1278.2, IEEE Standard for Distributed Interactive Simulation—Communication Services and Profiles, 1995.

[IJT94] Standard for Distributed Interactive Simulation–Application Protocols Version 2.0 Fourth Draft (Revised). Orlando: Institute for Simulation and Training, 1994.

[KNO98] Knoll F., "HORUS—Combat Simulation System," Ljubljana, 1998, No.98–4.

[KVA97] Kvamme, K., Oštir-Sedej, K., Stančič, Z., and Šumrada, R., "Geografski Informacijski Sistemi (Geographical Information Systems)," Ljubljana: Znanstve-noraziskovalni center Slovenske akademije znanosti in umetnosti, 1997, p. 108.

[KWA98] Kwat, J. D., "Synthetic Entities and Behavioral Representation," North Carolina State University, 1998.

[LAI94] Lai Y. J., and Hwang, C.L., "Fuzzy Multiple Objective Decision Making: Methods and Applications," Berlin: Springer-Verlag, 1994.

[LAW91] Law A. M., and Kelton W. D., "Simulation, Modelling and Analysis," New York: McGraw Hill, 1991.

[LIE84] Liebholz, S. W., "Twenty questions," In W. P. Huges J. (ed.), Military Modeling, (pp. 344–343), Arlington: Military Operations Research Society, 1984.

[LOC95] Locke, J., An Introduction to the Internet Networking Environment and SIMNET/DIS. Monterey: Computer Science Department, Naval Postgraduate School, 1995.

[MAC96] Mackinley, J., "A Guide to Peace Support Operation," Providence: Institute for International Studies, 1996, p. 24.

[MAR02] Marinčič, D., "CAX as a Training Method of the Peace Forces Components for the Peace Operations," Ljubljana: Faculty for Social Science, 2002, p. 102.

[MAR05] Marinčič, D., "Simulation and Analysis of Peace Operations," Ljubljana: Faculty for Social Science, Dissertation, 2005, p. 45.

[MAR79] Marušič, I., "Načrtovalska Analiza in Valorizacija Krajine (Planning Analysis and Valorization of Landscape)," Ljubljana: Biotehnična fakulteta. Agronomski oddelek.Katedra za krajinsko arhitekturo, 1979, pp. 2, 8, 12, 60, 70.

[MCF04] Mc Farlane, D., "Distributed Simulation Guide," Canberra: Department of Defence, 2004, pp. 3–4, 47.

[MOL00] Møller, B., "The Concept of Security: The Pros and Cons of Expansion and Contraction," København: Copenhagen Peace Research Institute. Working Papers 26, 2000, p. 2.

[MOL07] Moller B., Lofstrand B., and Karlsson M., "An Overview of the HLA Evolved Modular FOMs," Proceedings of Spring Simulation Interoperability Workshop, March 2007.

[NAR93] Nardi, B. A., "A Small Matter of Programming: Perspectives on End User Computing," Cambridge: MIT Press, 1993, p. 38.

[NAT06] National Defense and Canadian Forces, "Introduction," In Shaping the Future of the Canadian Forces: A Strategy for 2020, p. 6.

[NAT07] NATO., "Bi-Sc Exercise Directive 75–3," Mons: SHAPE, Norfolk: Supreme Allied Commander, Transformation, 2007, pp.1–1, 1–9.

[NIL02] Nilsson, J., Magnusson, S. E., Hallin, P. O., and Lenntorp, B., "Vulnerability Analysis and Auditing of Municipalities," Lucram: Lund University, 2002, p. 7.

[NOR06] North Atlantic Military Committee, "MC 0458/1 Education, Training, Exercise and Evaluation Policy," Brussels: International Military Staff, 2006, pp. 2–4, 3–4.

[PER90] Perla, P. P., "The Art of Wargaming," Annapolis: United States Naval Institute, 1990, p. 165, 274.

[PER06] Perry, N., "Verification and Validation of the Fractal Attrition Equation," Russell: DSTO. Defence Science and Technology Organization, Australia, 2006, pp. 2–3.

[POP02] Popper, K., "The Logic of Scientific Discovery," London: Routledge Classics, 2002, p. 139.

[PRE02] Prezelj, I., "Kompleksno ogrožanje varnosti in nastajanje kompleksnih razmer (Complex security threats and origins of complex circumstances)," In Malešič, M. (ed.) National and Imternational Security, Ljubljana: Faculty for Social Science, 2002, pp. 61–62.

[RAM00] Ramsbotham, O., "Reflections on UN post-settlement peacebuilding," In Ramsbotham, O., and Woodhouse, T., (eds.) "Peacekeeping and Conflict Resolution," London: Frank Cass Publishers Ltd., 2000, p. 171.

[SAV98] Savšek, T., "Odločanje na Podlagi Primerjave Mehkih Dreves (Decision Making on the Basis of Comparison of Soft Trees)," Doctoral dissertation. Ljubljana: Faculty of Electrotechnik, 1998, pp. LII, LIV, LIX, LXI, LXVII, LXVIII.

[SAV00] Savšek, T., "Sodobni Vojaški Simulacijski Sistemi (Modern military simulation systems)," Ljubljana: Ministry of Defense, 2000, pp. 27, 71, 72, 116, 148, 175, 184, 206.

[SCH74] Schellenberger, R. E., "Criteria for assessing model validity for managerial purposes," Decision Sciences J., Atlanta: Decision Sciences Institute, 1974, No. 5, pp.644–645.

[SIS98] Simulation Interoperability Standards Organization (SISO) Standards Development Group (SDG), Draft Standard [for] Modelling and Simulation (M&S) High Level Architecture (HLA)—Framework and Rules, IEEE P1516/D1. New York: Institute for Electrical and Electronic Engine, Inc., 1998.

[SMI06] Smith, E. A., "Complexity, Networking and Effects Based Approaches to Operations," Washington, DC: CCRP, 2006, pp. ix, xii.

[STE93] Stewart, B. D., "An Interactive Use of the Lancaster Combat Model," 1993.

[UNO07] UNODC, "Afghanistan opium winter rapid assessment survey," Vienna: UN Office on Drugs and Crime. Vienna International Centre, 2007, p. 9.

[VES91] Vester, F., Kriza Prenaseljenih Območij: o Razvijanju Ekosistemskega Mišljenja." Ljubljana: Državna založba Slovenije, 1991, p. 135.

[VIO94] Viotti, P. R., in Murray D. J., "The Defense Policies of Nations: A Comparative Study." Third edition. Baltimore: The John Hopkins University Press. 1994, p. 3.

[VRI82] Vrišer, I., "Uvod v Geografijo (Introduction in Geography)," Ljubljana: Filozofska fakulteta Univerze Edvarda Kardelja v Ljubljani. Oddelek za geografijo (Department for Geography), 1982, p. 414.

[WAG74] Wagner, R. A., and Fischer, M. J., "The string-to-string correction problem," J. Asso Computer Machinery, Vol. 21., No. 1, 1974, pp. 168–173.

[WAL87] Waltz E. L., and Buede, D. M., "Data Fusion and Decision Making for Command and Control," Principles of Command and Control, Washington, DC: AFCEA International Press, pp. 213–236.

INTERNET REFERENCES

[BAK07] Dr. Michael Baker website, http://www.michael-baker.com/cv/simtect-paper-1999.html, Baker, M., Payneham SA: Adacel Technologies Limited, "Time management in distributed simulation models," 2007.

[BRA98] Science Applications International Corporation website, http://people.cecs.ucf.edu/proctor/InteroperabilityText/Chapter3/SyntheticNaturalEnvFinal.htm, Braudaway, W., North Carolina State University, "Synthetic natural environments representation," 1998.

[BYR97] Sociological Research Online website, http://www.socresonline.org.uk/socresonline/2/2/4.html, Byrne, D., Durham, University of Durham, Sociological Research Online, No. 2, "Simulation—a way forward?," 1997.

[CSI02] Centre for Strategic & International Studies website, http://www.csis.org/media/csis/pubs/framework.pdf, Centre for Strategic and International Studies (CSIS) and the Association of the United States Army (AUSA), "Post-conflict reconstruction," 2002, p. 3.

[COB96] Aetheling Consultants website, http://www.aetheling.com/docs/Minimal.html, Cobb, L., Louisville, "Minimalist modeling methodology," 1996.

Dahmann, J., Evolution of DoD Mes strategy, Defense Modelling and Simulation Office, Alexandria. http://nla.dmso.mil/nla/general/annotate/index.htm, 1998.

[FIS02] University of Surrey website, http://www.soc.surrey.ac.uk/research/simsoc/bibliog.html, Fischer, D., and Guildford, D. M., University of Surrey, "Computer-based simulation modeling for anthropologists," 2002.

[GAT06] Air War College website, http://www.au.af.mil/au/awc/awcgate/navy/wft-a96.htm, Gattuso, J., Newport: Naval War College Review, "Warfare theory," 1996.

[HII07] Heidelberg Institute for International Conflict Research website, http://www.hiik.de/en/konfliktbarometer/pdf/ConflictBarometer_2008.pdf, 2008.

[HEN98] Parameters US Army War College Quarterly website, http://www.carlisle.army.mil/usawc/parameters/98autumn/henry.htm, Henry Ryan and C Edward Peartree, Washington, DC: Parameters, Autumn 1998, Vol. XXVIII, No. 3, "Military Theory and Information Warfare," 1998.

[ISB05] VIKING 05 website, http://www.mil.se/viking05/article.php?id = 14904, Isberg, J. G., "Briefing on viking 05, a computer assisted and distributed exercise," Stockholm: Press Centre, 2005.

[LEA98] Joint Electronic Library website, http://www.dtic.mil/doctrine/jel/jfq_pubs/1817pgs.pdf, Leahy, P., "ANZUS: A view from trenches," J. Joint Force Quarterly, Autumn/Winter 1997/1998 pp. 86–90, 1998. p. 87.

[MIS01] Ludwig von Mises Institute website, http://www.mises.org/nsande/pt2ch4.asp, Mises, Ludwig, IWDC. Inc., "The economy's war costs and the inflation," 2001.

[SED07] SEDRIS Technologies website, http://www.sedris.org/edcs.htm, The SEDRIS Organization, "The source for environmental data representation & interchange," 2007.

[SMI02] The Command and Control Research Program website, http://www.dodccrp.org/files/Smith_EBO.pdf, Smith Edward, Washington DC: CCRP Publication Series, Information Age Transformation Series, 2002, p. 92.

[SMI98] Watson Institute for International Studies Brown University website, http://www.watsoninstitute.org/pub/OP33.pdf, Smile Ian, Providence: The Thomas J. Watson Jr. Institute for International Studies Brown University, Occasional Paper # 33, "Relief and Development: The Struggle for Synergy," 1998, pp. xiv, xv.

[VLA95] The Journal of Humanitarian Assistance website, http://www.jha.ac/articles, Vlassenroot, R. D. K., Brussels: University of Ghent, "Early warning and conflict prevention: Minerva's wisdom," 1995.

ACRONYMS

AAR After Action Review
ACC Air Component Command
AF Automated Forces
AI Artificial Intelligence
ALSP Aggregate Level Simulation Protocol
API Application Program Interface
ATKIS Authoritative Topographic Cartographic Information System
ATO Air Tasking Order
BOM Base Object Model
BPA Battlefield Psychological Activities
C2 Command and Control Systems
C3 Command, Control and Communication Systems
C4I Command, Control, Communication, Computers and Intelligence
CAOC Combined Air Operation Centre
CAP Combat Air Patrol
CAX Computer-Assisted Exercise
CEO Complex Emergency Operation
CF Canadian Forces
CGF Computer Generated Forces

CIMIC Civil military Cooperation
CIMOC Civil military Operating Centre
CIS Communications and information system
CJTF Combined Joint Task Force
CMR Civil Military Relations
CMX Crises Management Exercise
CoA Course of Action
CONOPS Concept of Operations
CORBA Common Object Request Broker
CPT Core Planning Team
CPX Command Post Exercise
CRC Central RTI Component
CRO Crisis Response Operation
CRP Crisis Response Planning
DARPA Defense Advanced Research Project Agency
DBMS Database Management Systems
DDM Data Distribution Management
DDS Database Development System
DECT Digital Enhanced Cordless Telecommunications
DFAD Digital Feature Analysis Data
DHM Digital Height Model
DIF Data Interchange Formats
DIS Distributed Interactive Simulation
DLC API Dynamic Link Compatible Application Protocol Interface
DMA Defense Mapping Agency
DMSO Defense Modeling and Simulation Office
DMT Database Management Team
DOD Department of Defense
DRM Digital Relief Model
DTED Digital Terrain Elevation Data
DV-Day Distinguished Visitor Day
EBAO Effect-Based Approach to Operations
EDCS Environmental Data Coding Specification
EM Event managers
ENDEX End of the Exercise
EO Exercise Objectives
EP Exercise Process
EPG Exercise Planning Group
ESC Exercise Specification Conference

EU European Union
EXCEN Exercise Center
EXCON Exercise Control
EXDIR Exercise Director
EXPLAN Exercise Plan
EXSPEC Exercise Specifications
FAS Functional Area Systems
FCC Final Coordination Conference
FDD Federation Document Data
FEDEP Federation Development and Execution Process
FER Final Exercise Report
FGC Force Generation Conference
FIR First Impressions Reports
FOM Federation Object Model
FRAGO Fragmentary Order
GeoTIFF Geographic Tagged Image File Format
GIS Geographical Information System
GO Governmental Organization
HICON Higher Control
HLA High-Level Architecture
HNS Host Nation Support
HQ Headquarters
IC International Capabilities
IC International Community
ID Initiating Directive
IEEE The Institute of Electrical and Electronics Engineers
INFO OPS Information Operations
IO International Organization
IPC Initial Planning Conference
IS Information Systems
ISG International Solutions Group
IST Institute for Simulation and Training
ITD Interim Terrain Data
JCATS Joint Conflict and Tactical Simulation
JCO Joint Coordination Order
JDK Java Development Kit
JEC Joint Exercise Control
JEMM Joint Exercise Management Module
JEST Joint Exercise Scenario Tool

JFC Joint Force Command
JLCTC Joint Land Component Constructive Training Capability
JLVC Joint Live Virtual Constructive
JMRM Joint Multi Resolution Model
JNEM Joint Non-kinetic Effects Model
JOA Joint Operations Area
JTC Joint Training Confederation
JTLS Joint Theatre Level Simulation
LC Local Capabilities
LCC Land Component Command
LCG Linear Congruential Generators
LO Liaison Officer
LOCON Lower Control
LOGFED Logistic Federate
LRC Local RTI Component
LVC Live Virtual Constructive
MCC Maritime Component Command
MCDM Multi criteria Decision-making
MEL Master Environmental Library
MEL/MIL Master Events List/Master Incidents List
MOM Management Object Model
MOU Memorandum of Understanding
MPC Main Planning Conference
MRE Mission Rehearsal Exercises
MRM Machine Readable Media
MRT Mission Rehearsal Trainings
MSEL Master Scenario Event List
NAC North Atlantic Council
NATO North Atlantic Treaty Organization
NCW Network Centric Warfare
NGO Nongovernmental Organization
NIC Network Interface Cards
NIMA National Imagery and Mapping Agency
NTF NATO Training Federation
O/T Observer/Trainer
OCE Officer Conducting the Exercise
ODE Officer Directing Exercise
OMDT Object Model Development Tool
OMG Object Management Group

OMT Object Model Template
OODA Observe Orient Decide and Act
OP Operational Planning
OPLAN Operational Plan
OPP Operational Planning Process
OPR Officer of Primary Responsibility
ORBAT Order of Battle
OS Operating Systems
OSCE Organization for Security and Cooperation in Europe
OSE Officer Scheduling the Exercise
P2SN Persistent Partnership for Peace Simulation Network
PACS Personal Access Telecommunication System
PCA Psychological Consolidation Activities
PDG Product Development Group
PDU Protocol Data Units
PfP Partnership for Peace
PHS Personal Handy Phone System
PI Public Information
PIC Public Information Centre
POCC Psychological Operations Component Command
PSO Peace Support Operation
PSPA Peace Support Psychological Activities
PSYOPS Psychological Operations
PTA Primary Training Audience
RC Response Cell
RFI Request for Information
RLS Real Life Support
RMA Revolutions in Military Affairs
RMI Remote Method Invocation
ROE Rules of Engagement
RTI Run Time Infrastructure
SAF Semi-Automated Forces
SAT System Approach to Training
SC Security Council
SDCS SEDRIS Data Coding Standard
SDRM SEDRIS Data Representation Model
SE Synthetic Entities
SEAS Synthetic Environment for Analysis and Simulations
SEC Site Exercise Control

SEDEP Synthetic Environment Development and Exploitation Process
SEDRIS Source for Environmental Data Representation & Interchange
SIMNET Simulator networking
SISO Simulation Interoperability Standards Organization
SITCEN Situational center
SITFOR Situational Forces
SMA Strategic Military Assessment
SME Subject Matter Expert
SNE Synthetic Natural Environment
SOCC Special Operations Component Command
SOF Special Operation Forces
SOM Simulation Object Model
SOP Standing Operating Procedures
SPA Strategic Psychological Activities
SPG Strategic Planning Guidance
SR Societal Reconstruction
SRM Spatial Reference Model
STARTEX Start of the Exercise
SYNEX Synthetic Exercise
TA Training Audience
TDMT Terrain Data Management Tool
TEREDIT Terrain editor
TMU Terrain Modification Unit
TO Training Objective
TSO Timestamp Order
TT Training team
TTD Tactical Terrain Data
UN United Nations
UNODC UN Office on Drugs and Crime
UNSC United Nations Security Council
UPS Uninterruptable Power Supplies
VBS2 Virtual Battle Space 2
VMAP Vector Map
VOB Visitors' and Observers Bureau
VTC Video Teleconference
VV&A Verification, Validation and Accreditation
WC White Cell

INDEX

Addition rule, 73
Aggregate level simulation protocol (ALSP), 124
Ancillary tools, 245
Attribute, 134

Base object model (BOM), 140
Bayes' theorem, 74, 75

Central RTI component (CRC), 129
Checkpoint, 262
Civil-military cooperation, 6, 9, 37, 47–49, 158, 165
Combat modeling, 189, 192, 203
Combination, 67
Common object request broker (CORBA), 124
Communications and information systems, 40, 162, 268, 273, 256
Complex emergency operations, 5, 6, 37, 45–46
Complex emergency, 6, 7, 26, 28, 37, 44, 49, 55
Comprehensive approach, 16, 44, 51, 55
Computer simulation modeling, 11, 14, 176, 190, 196
Conditional probability, 70
Confidence interval, 89
Conflict, 19, 20–21, 27, 48
Contemporary security, 3, 36, 47, 176
Cool down period, 118
Counting
  techniques, 66
  rule, 66
Crash recovery, 262
Crisis response operations, 9, 33–35, 161, 164, 207, 210, 214
Critical values
  for the standard normal distribution, 97
  for the t distribution, 98
  for the χ^2 distribution, 99

Data array, 59–60
Dead reckoning, 125, 127
Decision making process, 12, 37, 44, 46, 206
Declaration management, 131
De Morgan's law, 74
Deviation, 63
Distributed exercise, 155–156
Distributed interactive simulation (DIS), 123–124, 155–157, 166

Distribution
binomial, 79, 88
continuous, 82, 91
discrete, 76, 88
exponential, 85, 91
Gaussian, 86
geometric, 79, 88
hypergeometric, 81, 88
negative binomial, 80, 88
normal, 86, 91
Poisson, 81, 88
rectangular, 85
uniform, 79
Dynamic link compatible (DLC), 128

Effect-based approach to operations, 43
Exercise, 7– 8, 27, 160, 163, 171–172, 175
centers, 267, 269,
command post, 9, 233
computer assisted, 10–17, 19, 155, 158, 160–161, 164, 171, 174, 181, 183, 185, 187, 196, 201, 207–208, 210, 214, 267–268, 272
control (EXCON), 11, 233
live, 9
management tool, 247
planning group (EPG), 11, 233
Experimentation, 143–145, 147–149, 151, 230, 268

Federation, 243
development and execution process (FEDEP), 137
management, 130
object model (FOM), 132
modular FOM, 137
Frequency, 60–61

Game crash, 262
Geographic information system, 196
Ghost entity, 127
Goodness of fit, 96

Higher level command (HICON), 11
High level architecture (HLA), 123, 128–129, 155, 157, 162
rules, 139
Histogram, 59
Hypothesis test, 92

IEEE 126, 128–129, 1278, 1516
Independent event, 70
Information operations, 40, 41
International Community, 3–5, 14, 34, 36, 44–45, 48–49, 176
International organizations, 4, 6, 40, 177

Joint exercise management module (JEMM), 248
Joint exercise scenario tool (JEST), 250

Linear congruential generator (LCG), 107
mixed, 107
multiplicative, 107
Local RTI component (LRC), 129, 135
Lookahead, 131
Lower level command (LOCON), 11
Logistics, 39–40

Main event list/main incident list (MEL/MIL), 177–178, 218, 267, 271
Mean, 62–63
Median, 62–63
Mediation-ware, 252
Mid square method, 106
Mode, 62–63
Multiplication rule, 70
Mutually exclusive event, 73

NATO training federation (NTF), 133
Network-Centric warfare, 38, 192
Nonkinetic warfare, 16
Nongovernmental organizations, 4, 6, 34–35, 40, 207
North Atlantic Treaty Organization (NATO), 9, 33, 161, 164–165

Object management, 131
Object model template (OMT), 128, 132
Operations, 27, 28, 31–32, 39
Ownership management, 131

Parameter, 135
Peace support operations, 35–36, 40, 46, 164
Percentile, 59
Permutation, 67
Probability, 64
Protocol data unit (PDU), 125–126
Pseudorandom number, 105
Psychological operations, 41–42

Random
variable, 76
number generation, 105
number period, 106

Range, 63
Remote method invocation, 124
Run ahead game, 263
Run time infrastructure (RTI), 128
  central RTI component (CRC), 129
  local RTI component (LRC), 129

Sample, 58
  of convenience, 58
  judgment, 58
Sampling
  random, 58
  systematic, 58
  stratified, 58
Scenario management tool, 249
Security, 260
Shadow game, 263
Simulation, 12–13, 16, 103, 149, 156, 161, 178, 190–191, 197, 206
  based acquisition, 103
  constructive, 13, 104, 234
  high resolution constructive, 13, 235
  highly aggregated constructive, 13, 239
  continuous, 116
  deterministic, 104
  discrete event, 116
  distributed, 16, 121
  dynamic, 104, 115
  live, 12, 104, 234
  military, 12
  model, 11, 156, 189–190, 194, 201
  Monte Carlo, 111
  non-kinetic warfare, 242
  object model (SOM), 132
  phases, 117
  static, 104, 111
  stochastic, 104
  virtual, 12, 104
Societal analysis, 7, 47
Societal security dimensions, 4, 7–8, 26
Softening, 127
Statistics, 58
  descriptive, 58–59
  inferential, 59, 89
Stem and leaf, 59
Strata, 58
System approach to training, 9–10,

Terrain modeling, 193
Time management, 131
Training audience (TA), 11
Training center, 267

Variance, 63

War, 20, 22–27, 29, 38
Warm-up period, 118

Printed and bound by CPI Group (UK) Ltd, Croydon, CR0 4YY
16/09/2026
14963409-0003